Modifications in Biomacromolecules

Edited by Xianquan Zhan and Atena Jabbari

Published in London, United Kingdom

Modifications in Biomacromolecules
http://dx.doi.org/10.5772/intechopen.97318
Edited by Xianquan Zhan and Atena Jabbari

Contributors
Kantharaju Kamanna, Aravind Dayanand Kamath, Khemchand R Surana, Ritesh B. Pawar, Ritesh A. Khairnar, Sunil K. Mahajan, Pradip Vitthal Hirapure, Jayisha Dhargawe, Rita Narayan Lakkakul, Yihua Gu, Shuangjie Wang, Yadong Li, Huijuan Shi, Yang Yang, Sheng-Ce Tao, Aijie Xin, Julius T. T. Dongdem, Simon P. Dawson, Robert Layfield, Shoji Mano, Kazumi Hikino, Masatake Kanai

First published in London, United Kingdom, 2024 by IntechOpen
IntechOpen is the global imprint of INTECHOPEN LIMITED, registered in England and Wales, registration number: 11086078, 5 Princes Gate Court, London, SW7 2QJ, United Kingdom
Printed in Croatia

British Library Cataloguing-in-Publication Data
A catalogue record for this book is available from the British Library

Additional hard and PDF copies can be obtained from orders@intechopen.com

Modifications in Biomacromolecules
Edited by Xianquan Zhan and Atena Jabbari
p. cm.

This title is part of the Biochemistry Book Series, Volume 51
Topic: Cell and Molecular Biology
Series Editor: Miroslav Blumenberg
Topic Editor: Rosa María Martínez-Espinosa

Print ISBN 978-1-80355-996-4
Online ISBN 978-1-80355-997-1
eBook (PDF) ISBN 978-1-80355-998-8
ISSN 2632-0983

IntechOpen Book Series
Biochemistry
Volume 51

Aims and Scope of the Series

Biochemistry, the study of chemical transformations occurring within living organisms, impacts all of the life sciences, from molecular crystallography and genetics, to ecology, medicine and population biology. Biochemistry studies macromolecules - proteins, nucleic acids, carbohydrates and lipids –their building blocks, structures, functions and interactions. Much of biochemistry is devoted to enzymes, proteins that catalyze chemical reactions, enzyme structures, mechanisms of action and their roles within cells. Biochemistry also studies small signaling molecules, coenzymes, inhibitors, vitamins and hormones, which play roles in the life process. Biochemical experimentation, besides coopting the methods of classical chemistry, e.g., chromatography, adopted new techniques, e.g., X-ray diffraction, electron microscopy, NMR, radioisotopes, and developed sophisticated microbial genetic tools, e.g., auxotroph mutants and their revertants, fermentation, etc. More recently, biochemistry embraced the 'big data' omics systems. Initial biochemical studies have been exclusively analytic: dissecting, purifying and examining individual components of a biological system; in exemplary words of Efraim Racker, (1913 –1991) "Don't waste clean thinking on dirty enzymes." Today, however, biochemistry is becoming more agglomerative and comprehensive, setting out to integrate and describe fully a particular biological system. The 'big data' metabolomics can define the complement of small molecules, e.g., in a soil or biofilm sample; proteomics can distinguish all the proteins comprising e.g., serum; metagenomics can identify all the genes in a complex environment e.g., the bovine rumen.
This Biochemistry Series will address both the current research on biomolecules, and the emerging trends with great promise.

Meet the Series Editor

Miroslav Blumenberg, Ph.D., was born in Subotica and received his BSc in Belgrade, Yugoslavia. He completed his Ph.D. at MIT in Organic Chemistry; he followed up his Ph.D. with two postdoctoral study periods at Stanford University. Since 1983, he has been a faculty member of the RO Perelman Department of Dermatology, NYU School of Medicine, where he is codirector of a training grant in cutaneous biology. Dr. Blumenberg's research is focused on the epidermis, expression of keratin genes, transcription profiling, keratinocyte differentiation, inflammatory diseases and cancers, and most recently the effects of the microbiome on the skin. He has published more than 100 peer-reviewed research articles and graduated numerous Ph.D. and postdoctoral students.

Meet the Volume Editors

Xianquan Zhan received his MD and Ph.D. in Preventive Medicine at West China University of Medical Sciences. He received his post-doctoral training in oncology and cancer proteomics at Central South University, China, and University of Tennessee Health Science Center (UTHSC), USA. He worked at UTHSC and the Cleveland Clinic in 2001–2012 and achieved the rank of associate professor at UTHSC. Currently, he is a full professor at Central South University and Shandong First Medical University, both in China, and an advisor to MS/PhD students and postdoctoral fellows. He is also a fellow of the Royal Society of Medicine and European Association for Predictive Preventive Personalized Medicine (EPMA), a national representative of EPMA, and a member of the American Society of Clinical Oncology (ASCO) and the American Association for the Advancement of Sciences (AAAS). He is also editor in chief of the *International Journal of Chronic Diseases & Therapy*, an associate editor of *EPMA Journal, Frontiers in Endocrinology*, and *BMC Medical Genomics*, and a guest editor of *Mass Spectrometry Reviews, Frontiers in Endocrinology, EPMA Journal*, and *Oxidative Medicine and Cellular Longevity*. He has published more than 170 articles, 29 book chapters, 8 books, and 3 international patents in the field of clinical proteomics and biomarkers.

Dr. Atena Jabbari received her BS in Chemistry in 2006, MS and Ph.D. in Organic Chemistry in 2010 and 2016, respectively, from Ferdowsi University of Mashhad, Iran. She conducted postdoctoral research at the Department of Medicinal Chemistry, School of Pharmacy, Mashhad University of Medical Sciences, Iran, in 2018. Dr. Jabbari is currently an Assistant Project Scientist at the Department of Chemistry and Biochemistry, University of California, Los Angeles (UCLA), USA. Her research interests include the synthesis of complex small molecules with biological activity, total synthesis of natural products, enzyme inhibition, and targeted cancer therapy. Her research studies have focused on drug design, synthesis of complex small molecules as enzyme inhibitors, SAR (Structure-Activity Relationship) studies, and targeted cancer treatment. She serves as an editorial board member and reviewer for several journals.

Contents

Preface

The central dogma of molecular biology explains the flow of sequence information between information-carrying biomacromolecules (DNAs, RNAs, and proteins). Modifications are the important molecular events that occur in these biomacromolecules that significantly regulate the structures and functions of biomacromolecules and even the entire biological system. Modifications in biomacromolecules have been recognized to be extensively associated with different pathological and physiological conditions, including cancer, inflammatory disease, neurodegenerative disease, metabolic disease, and diabetes. Modifications in biomacromolecules are very complex and important factors in their diversity. There are several modifications occurring in DNAs, such as cytosine methylation and hydroxymethylation, and at least 20 post-translational modifications occurring in DNA-binding protein histone to regulate the structures and functions of DNAs. There are at least 170 post-transcriptional modifications occurring in RNAs to regulate the structures and functions of RNAs. These include 3- and 5-methylcytosines (m3C, m5C), N1- and N6-methyladenosines (m1A, m6A, m6Am), pseudouridine (Ψ), 5-hydroxymethylcytosine (hm5C), and 2′-O-methylation (Nm). In proteins, there are 400–600 post-translational modifications to regulate the structures and functions of proteins. These include phosphorylation, glycosylation, ubiquitylation, methylation, sumoylation, acetylation, sulfation, deamidation, nitration, nitrosylation, hydroxylation, succinylation, prenylation, myristoylation, and palmitoylation. Different modifications in biomacromolecules have different characteristics and methodologies of research. The development of genomics, transcriptomics, proteomics, and bioinformatics is driving large-scale qualitative and quantitative studies of modifications of sites and levels in biomacromolecules that further clarify the cellular signaling molecular mechanisms and functions in which modifications are involved. Moreover, there are antagonistic and synergistic effects between different modifications in biomacromolecules, which significantly complicate their biological effects. To date, the studies on biomacromolecular modifications (**Figure 1**) are lacking and insufficient. Thus, it is time to strengthen the studies of biomacromolecular modifications in breadth and depth.

This book focuses on the global identification and quantification of different biomacromolecular modifications and their functional roles and molecular mechanisms in different pathological and physiological conditions. Chapter 1 addresses the structure and functions of ubiquitin and ubiquitin-meditated ubiquitination. Chapter 2 discusses the functions and roles of ubiquitination on peroxisomal membranes for protein transport. Chapter 3 discusses the structure and functions of prenylated proteins, which are important for signaling transduction and drug development for cancer. Chapter 4 addresses protein prenylation and its extensive applications. Chapter 5 discusses the prenylation of natural products and their roles in the development of antibacterial and antitumor drugs. Chapter 6 discusses the impacts of glycosylations on sperm fertility, as well as the lectin array method and its potential application for sperm function assessment. Chapter 7 discusses epigenetic modifications that include DNA methylation and histone modifications (acetylation, methylation, phosphorylation, SUMOylation, and ubiquitination) and their effects on human health and plant health.

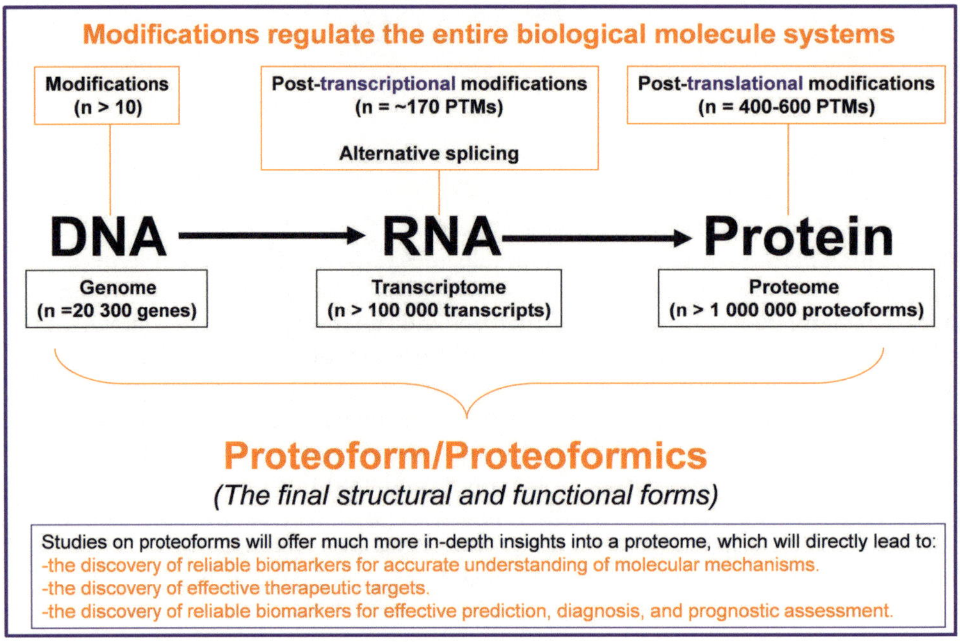

Figure 1.
Modifications regulate the structure and functions of biomacromolecules.

This book presents new advances and applications of modifications of biomacromolecules, including DNA methylation, histone modifications, ubiquitination, prenylation, and glycosylation. However, this book addresses only a very small fraction of biomacromolecular modifications, thus serving as a catalyst to stimulate and encourage researchers to conduct further biomacromolecular modification studies that result in important scientific developments in research and clinical practice of human health.

Xianquan Zhan, MD, Ph.D., FRSM
Professor of Cancer Proteomics and Predictive, Preventive, and Personalized Medicine (PPPM),
Medical Science and Technology Innovation Center,
Shandong First Medical University,
Shandong, China

Atena Jabbari
Assistant Project Scientist,
Department of Chemistry and Biochemistry,
University of California,
Los Angeles, USA

Chapter 1

Ubiquitin: Structure and Function

Julius T. Dongdem, Simon P. Dawson and Robert Layfield

Abstract

Ubiquitin is a small (8.6 kDa) protein that is found 'ubiquitously' in eukaryotic organisms and functions as a regulator of numerous cellular processes. It is a multifaceted post-translational modifier of other proteins involved in almost all eukaryotic biology. Once bound to a substrate, ubiquitin initiates a plethora of distinct signals with unique cellular outcomes known as the 'ubiquitin code'. More recently, much progress has been made in characterising the roles of distinct ubiquitin modifications though it is anticipated that more is yet to be unravelled as several questions remain elusive. The major aim of this chapter is to comprehensively review in detail using published data, the current understanding of the physico-chemical properties and structure (primary, secondary and tertiary) of ubiquitin, outlining current understanding of ubiquitin signal regulatory functions (Ubiquitin Proteasome System) and ubiquitin combinations, with emphasis on the structural relation to its function. Synthesis of ubiquitin (genes) will be illustrated. Additionally, ubiquitin-mediated processes and various possible covalent modifications of ubiquitin and their known functions will be illustrated. Deubiquitinase-dependent deubiquitylation of the ubiquitin code will also be described. Finally, ubiquitin-binding proteins and their ubiquitin-binding domains, the consequences of post-translational modification of ubiquitin by phosphorylation and future prospects will be discussed.

Keywords: gene, functions, ubiquitin, structure, modifications, ubiquitin-mediated processes, ubiquitin-binding domain, phosphorylated ubiquitin-binding protein

1. Introduction

Ubiquitin is a small (8.6 kDa) globular regulatory protein, which is found 'ubiquitously' in the cell-surface membrane, cytoplasm and nucleus of eukaryotic cells [1, 2]. Originally known as 'ubiquitous immunopoietic polypeptide' (UBIP), ubiquitin was first identified in calf thymus by Gideon Goldstein in 1975 in the search for thymopoietin and was further characterised through the 1980s [2–5]. Aaron Ciechanover, Avram Hershko and Irwin Rose first expounded the functions and the components of the ubiquitylation pathway, which earned the group the Nobel Prize for Chemistry, in 2004 [5–9]. The main function of ubiquitin is labelling of improperly folded, unwanted or damaged proteins for proteasomal degradation [10]. However, ubiquitin may also cause a change in the cellular location, structural conformation or biological function of other target protein substrates.

Understanding of ubiquitin signalling has broadened in recent years. Many novel proteins with non-covalent mono- and/or polyubiquitin binding activity have been discovered. These proteins, which are collectively referred to as ubiquitin-binding proteins (UBPs), contain ubiquitin-binding domains (UBDs, Section 7.0), which interact with ubiquitylated targets and regulate diverse biological processes such as endocytosis and DNA repair. This notwithstanding, the molecular mechanisms governing ubiquitin recognition in most cases have remained elusive [11, 12]. Dysfunction of ubiquitin signalling has been implicated in a wide range of diseases, including cancer, immune disorders, neurodegeneration, cardiovascular and metabolic disorders [13–15]. More recently, the demonstration that ubiquitin itself can be modified through phosphorylation by PINK1 provided a major breakthrough linking two very important signalling pathways in cells; phosphorylation and ubiquitylation [16–18]. The scope of ubiquitin signalling functions is probably much broader than first envisioned. The major objective of this chapter is to comprehensively review the current understanding of the structure of the ubiquitin protein, outlining ubiquitin signal regulatory functions, ubiquitin combinations and ubiquitin-mediated processes with emphasis on the structural relation to its function.

2. Physico-chemical properties of ubiquitin

Ubiquitin is a highly stable molecule over a wide range of pH and temperature values, exhibiting a melting point of ~100°C. Indeed, nuclear magnetic resonance (NMR) studies have shown that there is no significant denaturation of the ubiquitin fold over a temperature range of 23–80°C and a pH range of 1.18–8.48 [19]. The major contributor to ubiquitin stability is the large amount of intra-hydrogen bonding established within its entire structure, however, there are no disulphide bonds, coordinated metal ions or binding cofactors. Interestingly, if the ubiquitin protein is chemically denatured, the molecule can actually refold reversibly *in vitro* [20–22]. Ubiquitin is very resistant to tryptic (protease) digestion [2]. Only the two terminals -Gly^{75}-Gly^{76} are lost during proteolysis. This phenomenon is utilised in liquid chromatography with tandem mass spectrometry (LC-MS/MS) identification of covalent target conjugates of ubiquitin. The ubiquitin protein is composed of all common amino acids in different proportions except Trp. Its molar extinction coefficient [$\varepsilon\lambda$ = 280 nm] is 1280 [L/(mmol·cm)]. It has an Eetinction coefficient A280 of 1 = 6.69 mg/mL and an isoelectric point (pI) of 6.79. Net electric charge of ubiquitin at pH = 7 is −0.14 [3, 23].

3. Ubiquitin genes

The human genome consists of several copies of the ubiquitin sequence as a multigene family. The human ubiquitin is encoded by four independent genes: *UBB*, *UBC*, *UBA52* and *RPS27A* [24]. *UBA52* (located on chromosome band 19p13.1-p12) and *RPS27a* (chromosome band 2p16) encode a single copy of ubiquitin molecule, which is respectively fused to the ribosomal protein (RP) L40 and S27a (**Figure 1**). *UBB* and *UBC* are polyubiquitin precursor genes, with *UBB* encoding three ubiquitin moieties and *UBC* encoding nine ubiquitin monomers linked head-to-tail in tandem repeats by isopeptide linkages and without any spacer (intron) sequences. The last

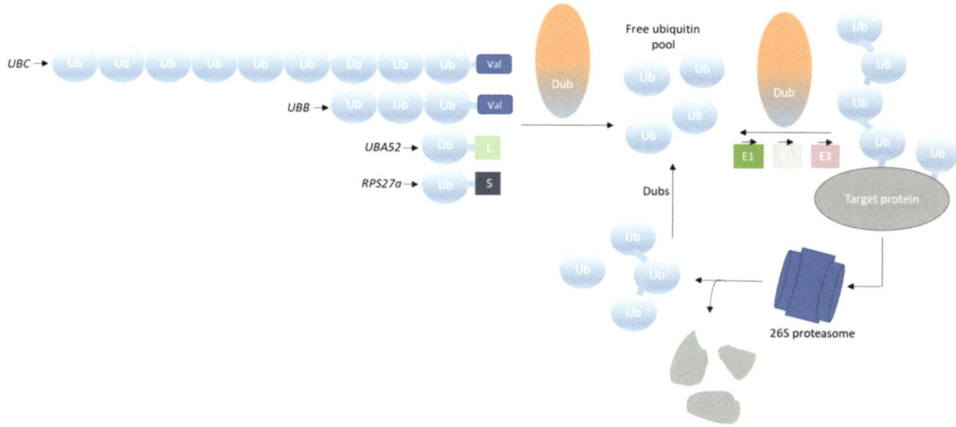

Figure 1.
Human ubiquitin genes and the ubiquitin-proteasome system (UPS). Ubiquitin is encoded by four independent genes in the human genome, including UBB, UBC, UBA52 and RPS27A. UBB encodes three ubiquitin moieties, while UBC encodes nine ubiquitin monomers linked head-to-tail in tandem repeats by isopeptide linkages, extended by an amino acid, for example, Val. UBA52 and RPS27a encode a single copy of ubiquitin molecule respectively fused to a RPL40 and RPS27a. This phenomenon is illustrated by the figure. It also illustrates how free ubiquitin molecules are generated by post-translational cleavage of precursor ubiquitin by dubs in order to maintain the ubiquitin pool. The 26S proteasome is responsible for degradation and recycling of unwanted proteins. Proteasomal proteolysis enables the cell to rid itself of these misfolded or damaged proteins and re-adjusts the concentration of essential proteins so that cellular homeostasis is maintained.

ubiquitin in the sequence is often extended with an additional Val residue in humans, and Tyr in chicken at the *C*-terminal end [25–28]. The number of tandem repeats of ubiquitin moieties varies depending on species and can also differ within one species. Species of kinetoplast eukaryotes, for example, *Trypanosoma cruzi*, contain polyubiquitin coding sequences with 52 tandem repeats of the ubiquitin gene. Generation of free ubiquitin molecules is by post-translational cleavage from the precursor ubiquitin moieties and is achieved by the action of deubiquitinases (Dubs), which hydrolyse the isopeptide bonds linking ubiquitin to the L40, S27a ribosomal proteins or amino acid linking the *UBB* or *UBC* translates (**Figure 1**) [29]. *UBB* gene and *UBB* pseudogenes are located on chromosome band 17p12-p11.1 while *UBC* is located on chromosome band 12q24.3 [25, 27, 30].

The physiological role of ubiquitin-coding genes as a source of ubiquitin remains largely elusive. *UBA52* is apparently essential in the development of embryo. It has been demonstrated that *UBA52* gene is not only a contributor to the ubiquitin pool, but also a regulator of the ribosomal protein complex (RPL40), downstream protein synthesis and cell-cycle arrest [31]. RPL40 is essential for translation of specific cellular transcripts. RPS27a is a regulator of microglia activation and is biologically relevant in triggering neurodegenerative diseases [32]. Other studies have underscored RPS27a as a signal transmitter between DNA damage response and cell cycle progression/ribosome biogenesis. RPS27a is implicated in the inhibition of apoptosis and promotion of cell proliferation [33, 34]. Both *RPS27a* and *UBA52* are preferentially over-expressed during hepatoma cell apoptosis [35]. An aberrant form of *UBB* translation has been detected in patients with Alzheimer's disease as well as patients with Down syndrome. Other diseases associated with *UBB* include cleft hard palate and submucosal cleft palate. *UBC* has been reported to be the most responsive gene to UV irradiation, heat shock, oxidative stress, proteotoxic stress and translational impairment, and is upregulated under these conditions [27, 35–38].

4. Structure of the ubiquitin protein

4.1 Primary structure

Ubiquitin is a 76-residue protein, which is evolutionarily one of the most highly conserved eukaryotic proteins known to date [39]. The amino acid sequence of the human ubiquitin for instance is 100% identical to that of sea slug aplysia. Similarly, primary sequences of ubiquitin isolated from bovine, fish, insects and humans are identical in the first 74 amino acids [40–43]. Yeast, barley, soya-bean, arabidopsis and oat ubiquitin primary sequences differ in only 3 out of 76 residues in comparison with that of higher eukaryotes [39]. Comparison of ubiquitin genes across eukaryotes have been extensively documented [44]. The amino acid sequence of a molecule of ubiquitin is shown in **Figure 2**. Ubiquitin possesses seven Lys residues (Lys^{6}, Lys^{11}, Lys^{27}, Lys^{29}, Lys^{33}, Lys^{48} and Lys^{63}), including an *N*-terminal Met$^{(1)}$ residue, which serves as linkage points for ubiquitylation in the formation of ubiquitin polymers. Each ubiquitin molecule also contains a *C*-terminal Gly^{76} whose free carboxylate group is used to bond with an ε-amino group of a substrate's Lys or that of preceding ubiquitins to generate various polyubiquitin chains (see later) [45].

4.2 Secondary structure

The most relevant secondary structural features found throughout ubiquitin include three helices; 3.5 turns of α-helix (15.8% of α-helix) and a short 3_{10} helix (7.9%) (**Table 1**; **Figure 3**). Each ubiquitin also contains one mixed β-sheet with five β-strands, no barrel and – 1 3X 1 –2X topology (**Table 2**) [46]. There are six β-reverse turns (**Table 3**) and two β-harpins whose characteristics are summarised in **Table 4** [47, 48]. Ubiquitin also contains a G1 type β-bulge, which is antiparallel (residue X: Tyr^{7}, residue 1: Gly^{10} and residue 2: Lys^{11}), two reverse Asx turns and a symmetrical hydrogen bonding region between two α-helices and two reverse turns (**Figure 3**) [46].

Met^{01}-Gln^{02}-Ile^{03}-Phe^{04}-Val^{05}-Lys^{06}-Thr^{07}-Leu^{08}-Thr^{09}-Gly^{10}-Lys^{11}-Thr^{12}-Ile^{13}-Thr^{14}-Leu^{15}-Glu^{16}-Val^{17}-Glu^{18}-Pro^{19}-Ser^{20}-Asp^{21}-Thr^{22}-Ile^{23}-Glu^{24}-Asn^{25}-Val^{26}-Lys^{27}-Ala^{28}-Lys^{29}-Ile^{30}-Gln^{31}-Asp^{32}-Lys^{33}-Glu^{34}-Gly^{35}-Ile^{36}-Pro^{37}-Pro^{38}-Asp^{39}-Gln^{40}-Gln^{41}-Arg^{42}-Leu^{43}-Ile^{44}-Phe^{45}-Ala^{46}-Gly^{47}-Lys^{48}-Gln^{49}-Leu^{50}-Glu^{51}-Asp^{52}-Gly^{53}-Arg^{54}-Thr^{55}-Leu^{56}-Ser^{57}-Asp^{58}-Tyr^{59}-Asn^{60}-Ile^{61}-Gln^{62}-Lys^{63}-Glu^{64}-Ser^{65}-Thr^{66}-Leu^{67}-His^{68}-Leu^{69}-Val^{70}-Leu^{71}-Arg^{72}-Leu^{73}-Arg^{74}-Gly^{75}-Gly^{76}

Figure 2.
Primary structure of the ubiquitin protein. The human ubiquitin amino acid sequence using their three-letter abbreviations is shown in the figure. Indicated are the seven Lys residues, including the four important C-terminal residues, respectively, highlighted in yellow and red. Positions of amino acids in the sequence are specified as superscripts.

Helix №	Start	End	Type	№ of residues	Length [Å]	Unit rise	Residues per turn	Pitch [Å]	Deviation [degrees]	Sequence
1	23	34	H	12	17.52	1.46	3.68	5.38	10.8	IENVKAKIQDKE
2	38	40	G	3	—	—	—	—	—	PDQ
3	56	59	G	4	6.77	1.69	3.45	5.82	40.1	LSDY

Table 1.
Secondary motives of the three helices in the human ubiquitin (15.8% of α-helix; 7.9% of 3_{10} helix).

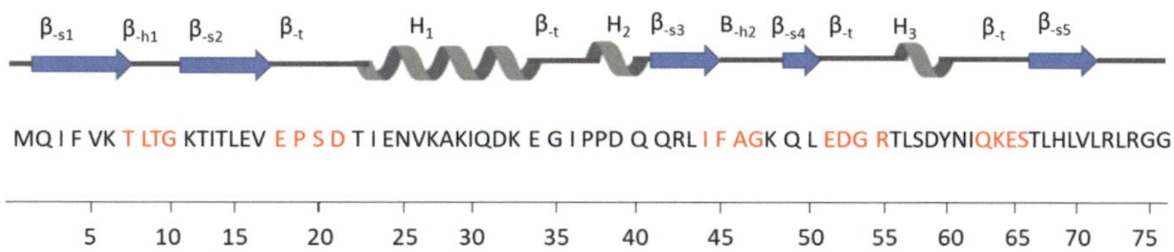

Figure 3.
A wiring illustration of the secondary structure of the human ubiquitin. The figure is an illustration of the most relevant secondary structural motives of ubiquitin stretched out from start to end in the form of a string. One-letter abbreviation of each amino acid has been mapped to the secondary structural motif below the structure. Amino acids that make up β-turns (β_{-t}) are highlighted in red colour. The five β-strands of ubiquitin ($\beta_{-s1\ to\ 5}$) are shown as blue sheets (arrows) while the three α-helices (H_1, H_2 and H_3) have been illustrated in grey. Positions of β-hairpins ($\beta_{-h1\ and\ 2}$) have also been indicated.

№ of stands	Start	End	Edge	№ of residues	Sequence
1	2	7	—	6	QIFVKT
2	12	16	+	5	TITLE
3	41	45	—	5	QRLIF
4	48	49	+	2	KQ
5	66	71	—	6	TLHLVL

Table 2.
Secondary motives of the five β-strands in the human ubiquitin (31.6%).

β-turn	Start	End	Turn type	H-bond	Sequence
1	7	10	I	+	TLTG
2	18	21	I	+	EPSD
3	44	47	IV	—	IFAG
4	45	48	I′	+	FAGK
5	51	54	I	+	EDGR
6	62	65	II	+	QEKS

Table 3.
Secondary motives of the six β-turns in the human ubiquitin.

Strand 1			Strand 2			Class
Start	End	Length	start	End	Length	
2	7	8	12	16	5	3:5
41	45	5	48	49	2	2:2

Table 4.
Two β-hairpins in the human ubiquitin.

4.3 Tertiary structure

Naturally, ubiquitin adopts a compact ß-grasp globular fold with a globular surface area of 4800 Å^2. It has been demonstrated that ubiquitin folds *via* a two-state process between a native and unfolded state in aqueous solution [49, 50]. Its terminal Gly^{76}-COOH freely protrudes from the globular structure allowing covalent modification of target proteins, including other ubiquitin moieties (**Figure 4**). Leu^8, Val^{70} and Ile^{44} residues of ubiquitin form a hydrophobic surface patch centred around Ile^{44}, which in combination with electrostatic potential resulting from positively charged residues, including Lys^6, Arg^{42}, Lys^{48}, His^{68} and Arg^{72}, are relevant for the non-covalent interaction with many UBDs [51, 52]. The hydrophobic patch also facilitates intra-chain interactions between certain polyubiquitin chains. Additional hydrophobic surfaces include a patch centred on Ile^{36} involving Leu^{71} and Leu^{73} as well as the Phe^4 patch, which involves Gln^2 and Thr^{12}. The 'TEK-box' of ubiquitin found in higher eukaryotes is a three-dimensional motif comprising Thr^{12}-Thr^{14}-Glu^{34}-Lys^6-Lys^{11} and is required

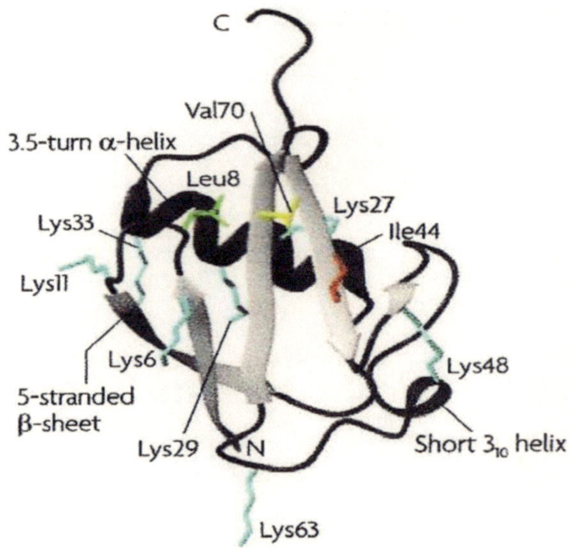

Figure 4.
Three-dimensional structure of ubiquitin. Three-dimensional structure of human ubiquitin indicating functionally relevant amino acids with exposed secondary structures; the seven Lys residues, including the Leu^8-Ile^{44}-Val^{70} hydrophobic patch. Hydrophobic patches serve as platforms for many UBDs. Side chains of Lys residues are highlighted in cyan. β-strands are shown in grey arrows while the 3.5-turn of α-helix and 3_{10} helix are in black. The C-terminal Gly^{76}-COOH is shown protruding at the top of the structure. N-terminal met is shown at the bottom of the structure (reproduced from Dikic et al. [51]).

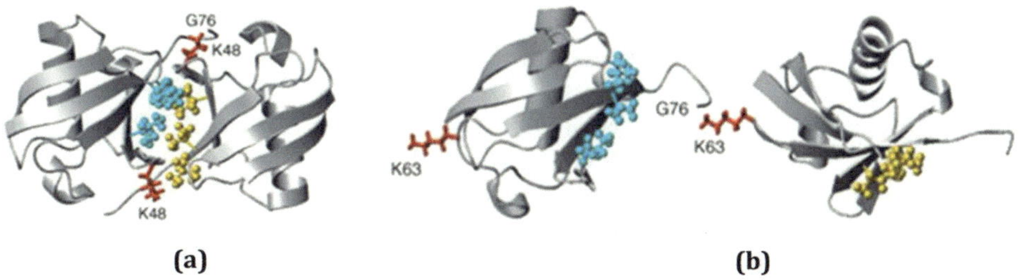

Figure 5.
Conformation of (a) Lys^{48}-linked diubiquitin and (b) Lys^{63}-linked diubiquitin. Hydrophobic patches are (recognised by UBDs) highlighted in gold and cyan in the ball and stick model. The side chains of Lys^{48} and Lys^{63} are highlighted in red (reproduced from Pickart and Fushman, [64]).

for mitotic degradation [53, 54]. Another interaction surface includes the *C*-terminal di-Gly^{75,76} motif of ubiquitin, which interacts with the ZnF (Zn finger)-UBP, the UBD of HDAC6 during aggresome formation (i.e. intracellular protein aggregation) [55]. Characterisation of a novel acidic interaction (hydrophilic) surface centred on Asp^{58} has been found to be recognised by the ZnF_A20 UBD of Rabex-5 [56, 57]. Ubiquitin exhibits several strategies that maintain its three-dimensional structure in a very stable state. There is almost no noticeable change in conformation of ubiquitin in water solution and in crystalline form. All three Pro residues$^{19, 37,38}$ of ubiquitin (at 1.8 Å crystal structure, 1UBQ. pdb) display the Cγc-exo conformation. Pro^{19} is a component of a flexible loop that connects the *N*-terminal β-harpin of ubiquitin to the 3.5 α-helix. Pro^{37} and Pro^{38} are, however, placed in an extended loop that links the *C*-terminal end of the 3.5 α-helix to the β-strands [58]. These pro residues are reported to play critical roles in the conformational stability of ubiquitin [59]. Vijay-Kumar et al. [2] determined the three-dimensional structure of ubiquitin at a resolution of 2.8 Å and stated that the His, Tyr and the two Phe residues are located on the surface of the molecule.

Ubiquitin usually exists either covalently attached to other proteins or another ubiquitin or free (unanchored), that is, covalent assemblies of multiple ubiquitins in a substrate-free form [60–62]. Lys^{48}-linked diubiquitin adopts a closed conformation with the hydrophobic residues (Leu^{8}, Ile^{44}, Val^{70}) forming the inter-domain interface [63]. Due to weak ubiquitin-ubiquitin interactions, the interface is not rigidly locked. This allows the functional hydrophobic residues to be accessible for interactions with various recognition domains (**Figure 5a**). Lys63-linked diubiquitin, however, adopts an extended conformation and therefore, no hydrophobic interaction exists between its ubiquitin units (**Figure 5b**).

5. Ubiquitin modifications and functions

5.1 Major functions

As aforementioned, the major function of ubiquitin is regulation of the degradation of other proteins [6, 7, 64]. Ubiquitin plays a role in the intracellular ATP-dependent, non-lysosomal proteolysis and elimination of defective proteins, normal proteins with a rapid turnover, as well as certain short-lived regulatory proteins within the cytoplasm. The mechanism of the 26S proteasomal degradation of unwanted proteins involves covalent binding of a polyubiquitin chain of about four or more ubiquitin moieties to the target proteins to be degraded by formation of isopeptide bonds between the free carboxylate group of ubiquitin's *C*-terminal Gly^{76} and an ε-amino group of the substrates' Lysyl side chains in a process known as ubiquitylation. Protein ubiquitylation is a post-translational modification (PTM) event carried out in a three-step enzymatic process by an E1 ubiquitin-activating enzyme, E2 ubiquitin-conjugating enzyme and E3 ubiquitin ligase [13, 28, 45]. Post-translational modifications (PTMs), which are usually covalent and reversible, may alter the properties and therefore, the functions of the modified protein. Proteasomal proteolysis enables the cell to rid itself of these unwanted, short-lived, inactive, oxidised, unfolded, misfolded or damaged proteins and re-adjusts the concentration of essential proteins so that cellular homeostasis is maintained (**Figure 1**). Protein ubiquitylation plays a plethora of roles aside from proteasomal degradation [45, 65]. The ubiquitin code is able to modulate cell cycle progression, cell differentiation, several signal transduction pathways and membrane protein trafficking. Ubiquitin induces differentiation of both B (bone-marrow-derived) and T (thymus-derived) lymphocytes [1]. Ubiquitin cross-reactive protein, a 15 kDa protein believed to be an example of functionally distinct family of ubiquitin-like proteins, is essential in various cellular responses to biological effects of interferons [66]. Strikingly, ubiquitin acts as an immunophilin, in that it is able to bind immunosuppressive drugs such as tacrolimus (FK506, Kd = 0.8 nM) and sirolimus (rapamycin, Kd = 0.08 nM). Ubiquitin complexes with tacrolimus, which acts as an inhibitor of calcineurin (protein phosphatase 2B) involved in T cell activation [67]. A peptide fragment of ubiquitin; Leu^{50}-Glu^{51}-Asp^{52}-Gly^{53}-Arg^{54}-Thr^{55}-Leu^{56}-Ser^{57}-Asp^{58}-Tyr^{59} possesses a very high immunosuppressive activity in both cellular and humoral immune responses that is at the level of cyclosporine, with Leu^{50}-Glu^{51}-Asp^{52}-Gly^{53}-Arg^{54}-Thr^{55}-Leu^{56} fragment been the shortest possible fragment with effective immunosuppressive activity [68]. Ubiquitin is a component of several cytoplasmic inclusions such as Rosenthal fibres, Mallory bodies, Crooke bodies, Lafora bodies and amyloid bodies, and of neurones such as the giant axonal neuropathy (GAN) and paired helical filaments (PHF), which constitute a distinct type of pathological neuronal fibre. These make up the principal constituent of neurofibrillary tangles (NFL) that occur in the brain of patients with Alzheimer's and Down disease [69, 70]. Apparently, association of ubiquitin with these cell inclusions may signify ATP-dependent proteolysis of these bodies, for example, NFL. Subcellular localisation of proteins where they control other protein functions and cell mechanisms are regulated by the ubiquitin code. For example, ubiquitin is involved in the ATP-dependent insertion of monoamine oxidase A and B into the outer mitochondrial membrane [71]. Autophagy, transcription, inflammatory signalling, modulation of enzymatic activity, DNA repair, stress responses, embryogenesis, cell apoptosis, virus budding, vacuolar protein sorting, inflammatory response and receptor endocytosis are regulated by ubiquitin-mediated signalling [64, 72–74]. Ubiquitin is responsible for the organisation and maintenance of chromatin structure by binding histone H2A in the nucleus. The first of such conjugations of ubiquitin described was binding histone H2A through an isopeptide bond with ε-NH_2- group of Lys^{119} of histone H2A [4, 75]. Additionally, ubiquitin regulates heat shock responses and is a constituent of certain cell surface receptors [76, 77]. Ubiquitin has also been demonstrated to play a role in the regulation of gene expression [78].

5.2 Ubiquitin-mediated processes

The cell must continually maintain its internal homeostatic conditions. The lifespan of each protein is highly regulated. The ubiquitin-proteasome system (UPS) is a major ubiquitin-mediated process recognised as the cellular protein quality control system in that it selectively targets all unwanted or damaged proteins, which would otherwise accumulate and destroy neurones among others for proteasomal degradation. The UPS, which involves a complex combination of several enzymes (over 1000 proteins in human ubiquitylation), is now one of the most important systems required for the regulation of protein function because it is involved in nearly all the important cell biological activities, such as cell metabolism, cell proliferation, glycogen synthesis and cell death, as well as in disease pathogenesis, for example, inflammation, arthritis, heart disease and cancers [79]. The UPS plays essential role in protein homeostasis in that the system regulates the turnover of proteins required for the plethora of regulatory pathways responsible for several cellular processes, for example, DNA damage and repair, cell cycle progression, apoptosis, etc. As such dysfunction of the ubiquitin pathway often results in pathological conditions.

Ubiquitin conjugated on target proteins can, however, be removed by Dubs into free ubiquitin or unanchored ubiquitin chains (**Figure 1**). Degradation of ubiquitin chains by the UPS machinery also recycles ubiquitin molecules to maintain homeostasis in the cells [80]. Dubs, therefore, further add complexity to the inherent ubiquitin code by acting as proofreading enzymes, which increase the specificity of the UPS (see Section 6.0). The occurrence of different ubiquitin-ubiquitin linkage types further complicates the process and also indicates other potential regulatory functions yet to be discovered.

5.3 Role of ubiquitin modules

Polyubiquitin chains, when attached to a target protein, exhibit different functions depending on the Lys residue of the ubiquitin involved in the formation of the linkage. This arises as a result of changes in shape of the polyubiquitin chain in which a unique and complex ubiquitin code is generated depending on the specific Lys residue on which the preceding ubiquitin molecule is covalently attached to the next ubiquitin molecule on the target protein [81]. The versatility of ubiquitin to potentially exhibit diverse and highly complicated linkage-specific type PTMs of target proteins is by virtue of the occurrence of an *N*-terminal Met $^{(1)}$ together with the seven Lys residues per ubiquitin moiety. Monoubiquitylation refers to the conjugation of a single ubiquitin molecule to a single Lys of the target protein (**Figure 6**). Multimonoubiquitylation implies that a target protein is tagged with more than one single molecule of ubiquitin. When the target protein is tagged with a polyubiquitin chain linked through the *C*-terminal Gly^{76} of each ubiquitin unit and a specific internal Lys of the previously attached ubiquitin through a series of ubiquitylation, the module is known as polyubiquitylation (**Figure 6**). A polyubiquitin chain is termed homogenous when the ubiquitin monomers are joined together through a single ubiquitin-ubiquitin linkage-type, whereas a heterogeneous polyubiquitin chain contains more than one single linkage type (**Figure 6**) [81]. Mixed polyubiquitin chains on the other hand contain one linkage type, which is extended by a second type and so on, forming a non-branched structure. However, a polyubiquitin chain is said to be branched when different linkage types form one or more branches linking multiple Lys residues in the same ubiquitin [82].

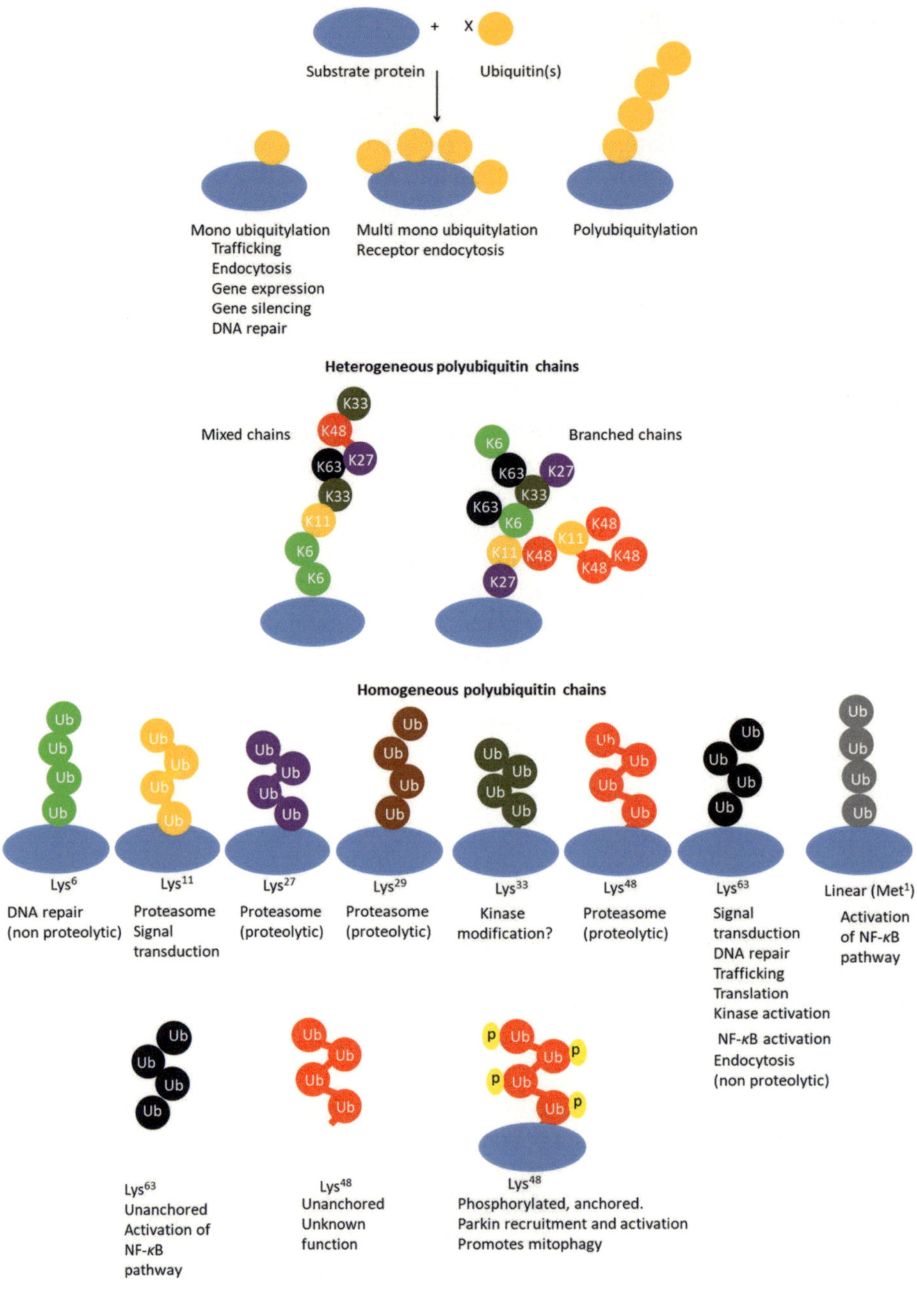

Figure 6.
Multifaceted ubiquitin modifications. The figure illustrates different architectural modules of ubiquitin-ubiquitin linkage and some of the known cellular processes they have been found to regulate to date.

Typically, both monoubiquitylation and multimonoubiquitylation on a single protein regulate intra- and inter-molecular interactions [81]. For instance, response to a genotoxic stress is mediated by monoubiquitylation of PCNA's Lys^{164} resulting in subsequent recruitment of DNA polymerases lacking proofreading activity to overcome a DNA replication block and bypass DNA lesions [83]. Additionally, monoubiquitylation has been implicated in the endocytic trafficking of certain cargo proteins, for example, small GTPases and receptors such as the EGFR to specific cellular compartments at different stages of the endocytic pathway. Monoubiquitylation is known to regulate histone modification [84]. Monoubiquitylation has also been implicated in the regulation of gene expression as well as DNA repair [85–88]. Multimonoubiquitylation is important for receptor endocytosis [89]. As noted previously, Lys^{48}-linked polyubiquitylation (i.e. polyubiquitin chains linked through Lys^{48} of the proximal ubiquitin to the next ubiquitin moiety in the chain) are the primary targeting signals for proteasomal degradation [90–92]. Lys^{63}-linked polyubiquitin chains function as scaffolds to assemble signalling complexes as in the activation of the transcription factor NF-κB, which is involved in inflammatory and immune response, DNA damage tolerance, the endocytic pathway and ribosomal protein synthesis [93]. Unanchored (substrate-free) polyubiquitin also has distinct signalling roles, including activation of protein kinases (**Figure 6**). Xia et al. [94] showed that unanchored Lys^{63}-linked polyubiquitin chain assembled through the action of UBE2N/UBE2V1 (an E2 conjugating enzyme) and TRAF6 (the E3 ligating enzyme) also activate the NF-κB pathway by activating the kinase TAK1, which in turn phosphorylates and activates IκB kinase (IKK). While Lys^{6}-linked ubiquitin chains are reported to be involved in DNA repair, Lys^{11}-linked polyubiquitin chains are involved in endoplasmic reticulum-associated degradation (ERAD) as well as the regulation of cell-cycle. Lys^{11}-linked polyubiquitin chains also target proteins for proteasomal degradation with Lys^{48}-linked polyubiquitin being the most efficient trigger of degradation [95]. Whereas Lys^{29}-linked polyubiquitylation is involved in proteotoxic stress response and cell cycle regulation, Lys^{33}-linked ubiquitylation is apparently required for kinase modification. Linear linkage *via* *N*-terminal Met is reported to regulate cell signalling such as NF-κB signalling [96–98].

Though the function of phospho-Ser65-(poly)ubiquitin in parkin activation has been established, the molecular consequences of ubiquitin phosphorylation have not been fully explored. Additionally, little is known about the existence and therefore functions of phosphorylated unanchored (poly)ubiquitin. Also, much less is known about more complicated polyubiquitin architectures, including heterogenous chains having branches or mixed chains [99].

6. Deubiquitylation

The action of the E1, E2 and E3 enzyme cascade in the process of ubiquitylation is antagonised by deubiquitylation proteases or Dubs. Dubs are proteases involved in the removal of ubiquitin from ubiquitylated proteins, as well as cleaving ubiquitin from its precursors [100]. Dubs cleave the isopeptide bonds between the tagged protein and ubiquitin or ubiquitin molecules. Immediately after a target protein has been delivered to the 26S proteasome, the tagged ubiquitin chain is cleaved off by proteasome-associated Dubs prior to target destruction after which the ubiquitin chain is likewise cleaved to produce free ubiquitin, which is recycled [101]. Generically, Dubs are capable of exo-activity (hydrolysis of ubiquitin-linkage from ends) and endo-activity (hydrolysis from within an ubiquitin polymer), however, in each case, an existing scissile linkage between an ε-NH_2 group of a Lys in the distal ubiquitin or substrate protein and the *C*-terminus of the proximal ubiquitin is cleaved.

Approximately, 100 putative Dubs have been characterised from the human genome and have been grouped into five distinct families based on the architecture of their catalytic domains [100]. They include ubiquitin *C*-terminal hydrolases (UCH), ubiquitin-specific proteases (USP), the ovarian tumour proteases (OTU) and the five Machado-Josephin domain proteases (MJDs/Josephins). These four subfamilies are Cys proteases, which rely on a conserved Asp/Asn, His and Cys catalytic triad mechanism of action. In contrast, the fifth subfamily, the Jun activating binding protein 1 (JAB1/MPN/MOV34) also known as JAMMs are Zn^{2+} metalloproteases, which invariably utilise Asp, His and Ser residues to coordinate a catalytic Zn^{2+} [102]. Dubs recognise (poly)ubiquitin non-covalently through intrinsic catalytic core domains or mediated by UBDs, including the ZnF-UBP domain, ubiquitin-interacting motif (UIM), ubiquitin-associated (UBA) domain and ubiquitin-like (Ubl) domain. Dubs, therefore, represent a specialised class of UBPs [80].

Dubs have been implicated in the regulation of various cellular events and have recently emerged as attractive therapeutic targets. They play several significant roles in the ubiquitin-mediated processes [103]. Dubs carry out processing of ubiquitin precursors, most probably co-translationally (**Figure 1**). As already mentioned, ubiquitin is expressed as a pro-protein fused to either ribosomal proteins or as linear polyubiquitin plus an additional amino acid and must, therefore, be hydrolysed to yield the free mature (poly)ubiquitin [28, 104, 105]. They also hydrolyse unanchored polyubiquitin released from target proteins or synthesised by *de novo* conjugating machinery into free ubiquitin molecules, for example, Ubp1 [106–108]. By so doing, Dubs maintain the homeostatic control of cellular flux of free ubiquitin levels [108, 109]. Dubs are also responsible for rescuing proteins from ubiquitin-mediated proteasomal degradation thereby stabilising the proteins. By antagonising the process of ubiquitylation, Dubs reverse the ubiquitin-mediated signalling cascades in response to cellular environmental changes [100, 103]. Dubs may also edit the form of ubiquitin modification by trimming polyubiquitin chains on substrate proteins [80, 102, 110].

Dubs not only frequently associate with E3 ligases but are themselves often modified by ubiquitin. This suggests that Dub activity may also be regulated by E3 ligases and *vice versa*. While the basic enzymatic function of Dubs is understood, how Dub activity is regulated, the cellular pathways that are regulated by Dub activity, the cellular target substrates of Dub regulation as well as how substrate recognition is regulated have not been fully explored. Notably, ubiquitin may become Dub-resistant once phosphorylated. Significant inhibition of the hydrolytic activity of Dub enzymes by phosphorylation of ubiquitin has been reported [111–115].

7. Ubiquitin-binding proteins and their ubiquitin-binding domains

A UBD is a modular structural motif, typically less than 50 residues of a protein that can bind non-covalently to a ubiquitin moiety. UBDs mediate the non-covalent recognition of (poly)ubiquitin modifications and interpret and transmit information conferred by protein ubiquitylation to control various cellular events.

Proteins that contain UBDs are collectively referred to as UBPs and are classified according to the type of UBD they possess. Most UBPs contain one or more recognisable UBD(s) (**Figure 7**). The majority of UBPs interact through ubiquitin's canonical Ile^{44}/Val^{70} hydrophobic patch on its β-sheet surface (**Figure 4**). A number of UBPs, however, bind to non-canonical sites such as the *C*-terminal di-Gly, the polar surface centred on

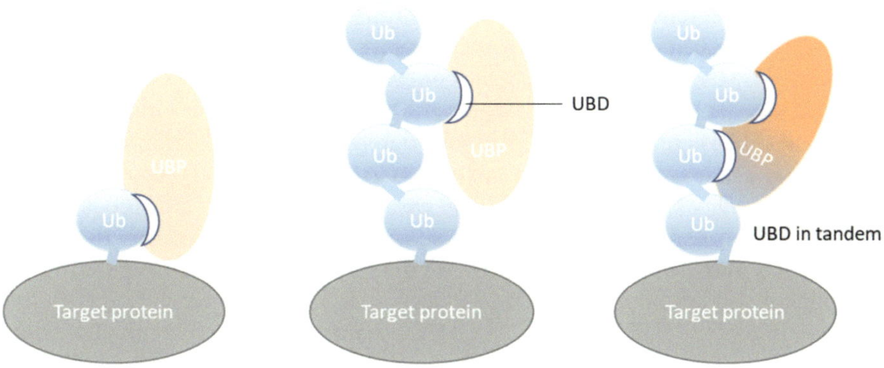

Figure 7.
Ubiquitin interactions with ubiquitin-binding proteins in the control of cellular dynamics. An illustration of UBPs which contain UBDs typically less than 50 residues in length. UBDs adopt many different folds and are capable of interacting non-covalently with both mono- (left panel) and polyubiquitylated (middle panel) targets. In so doing, UBPs interpret and transmit information conferred by covalent protein ubiquitylation to control various cellular events.

Asp^{58}, and the hydrophobic patch centred on Leu^{8} or Ile^{36} [55, 116]. Although individual UBDs bind ubiquitin with low affinity, they do so with high specificity, a probable requirement for rapid assembly and disassembly of ubiquitin–UBD complexes. To date, most isolated UBDs bind monoubiquitin, as such do not exhibit linkage specificity, and are involved in the regulation of a variety of cellular processes such as endocytosis, DNA repair, vesicular trafficking or signal transduction (**Figure 7**, left panel). A number of UBDs bind polyubiquitin and can differentiate the eight ubiquitin linkages, which are essential for linkage-selective functions (**Figure 7**, middle panel) [117]. Where two or more UBDs simultaneously recognise different ubiquitin moieties within a specific polyubiquitin chain *via* avidity effects, they typically align in tandem on the ubiquitin chain (**Figure 7**, right panel) [118]. Currently, more than 20 different families of UBDs have been characterised based on structure. Nevertheless, additional UBDs will undoubtedly be discovered [51, 118, 119].

7.1 Families of ubiquitin-binding proteins

Several different families of UBPs have been described based on the type of UBDs they possess. Proteins that contain UBDs include ubiquitylation enzymes and Dubs as well as ubiquitin receptors. Ubiquitin-receptors (non-catalytic effector proteins) carry UBDs, thus, they recognise and bind ubiquitylated proteins and subsequently decode information into a specific cellular response. The activity of some ubiquitin receptors is controlled by self-monoubiquitylation, referred to as coupled monoubiquitylation. Coupled monoubiquitylation has been shown to involve monoubiquitylation of a ubiquitin ligase and its subsequent interaction with a ubiquitin receptor. Coupled monoubiquitylation provides an efficient switch from an active to an inactive conformation *via* an auto-ubiquitylation mode where the conjugated ubiquitin masks the UBD sequence [120].

Some examples of UBPs include S5a/Rpn10/Pus1 and S6'/Rpt5, which are two proteasomal subunits identified to interact non-covalently with Lys^{48}-linked polyubiquitin chains of four or more moieties. The first UBP to be identified was S5a. The S5a/Rpn10/Pus1 ubiquitin-binding region was mapped to an approximate 20 residue region near the *C*-terminus. Mammalian S5a/Rpn10/Pus1 contains two of such regions, now recognised as the ubiquitin-interacting motif (UIM), and binds ubiquitin through hydrophobic interactions [121–123]. The Rpn13 protein also plays a role in proteasome function and is characterised by the Pleckstrin-like receptor for ubiquitin (PRU) domain [124]. Yeast protein Rad23/hHR23A (the human homologue) and Dsk2/Dhp1 are ubiquitin-associated domain (UBA) containing UBPs involved in shuttling ubiquitylated proteins to the proteasome whereas that of neighbour of BRCA1 gene 1 (NBR1) plays an important function in autophagy [125]. Whereas NF-$_k$ß essential modulator (NEMO), ABIN1-ABIN3 and OPTN are involved in NF-$_k$ß signalling and bind ubiquitin *via* their ubiquitin binding in ABIN and NEMO domain (UBAN) domain [126, 127], the Prp8 is involved in RNA splicing and employs its JAB1/MPN domain to bind ubiquitin [128]. UBCH5C (UBE2D3) contains the UBC domain. UBCH5C accepts ubiquitin from the E1 complex and catalyses its covalent attachment to other proteins [129]. Several other UBPs (e.g., NDP52, p62), their domain structure and functions have been extensively reviewed by many researchers [118, 130, 131].

7.2 Phosphorylated ubiquitin and phosphorylated ubiquitin-binding proteins

Eukaryotes have a large repertoire of PTMs, which include phosphorylation, acetylation, methylation, ubiquitylation, SUMOylation neddylation, glycosylation and glycation. PTMs act as a common mechanism for modulating and regulating protein function by considerably changing surfaces and binding properties of a protein. Phosphorylation is the biochemical process by which a phosphoryl (PO_3^{2-}) group is added to a protein. Protein phosphorylation often results in the alteration of the tertiary structure, function and activity of the modified protein. Ubiquitin function can in turn be modulated by PTMs such as acetylation (on Lys residues) or phosphorylation. Each of these modifications has potential to dramatically alter the fate or signalling outcome of the modified protein [115, 132–134]. Phosphorylation of ubiquitin on Thr^7, Thr^{12}, Thr^{14}, Ser^{20}, Thr^{22}, Thr^{55}, Ser^{57}, Ser^{65}, Thr^{66} and Tyr^{59} has been reported in various studies [133, 135, 136], however, the kinases that instal phosphorylations on these residues are largely unknown with the exception of PINK1 [18] and the molecular consequences of phosphorylations on these sites have not been fully characterised [132–140].

8. Conclusion and perspective

Although the ubiquitin structure has been resolved and is well documented, and recent studies have made creditable advances to understand the molecular basis of ubiquitin regulatory roles in higher-order polyubiquitin architecture, many questions have remained elusive. Much less is known about functions of more complicated polyubiquitin architectures such as heterogenous (branched and mixed) chains. The molecular mechanisms governing ubiquitin recognition by UBDs are not also well understood. It is believed that there are many more UBPs and UBDs than are currently known to exist. Additionally, little is known about the linkage type-specific functions of unanchored (poly)ubiquitin and phosphorylated unanchored polyubiquitin. The complex nature of ubiquitin modulation by phosphorylation has raised further questions. There are nine phosphorylatable Ser and Thr residues in ubiquitin besides Ser^{65}. The molecular significance of their phosphorylation has not been fully investigated. The kinases (besides PINK1, i.e., the Ser^{65}-ubiquitin kinase), phosphatases and selective recognition of ubiquitin targets induced by phosphorylations or dephosphorylations at these sites in brain and neurones have not been fully understood. How Dub activity is regulated, cellular pathways regulated by Dubs and how substrate recognition is regulated are yet to be explored. The current limitations preventing more insights into unanswered questions are probably due to lack of the most relevant and versatile technology for assessing these principles. However, it is envisioned that more insights regarding unanswered questions on ubiquitin biology will be unravelled in the near future.

Acknowledgements

We thank IntechOpen Ltd. UK, for an opportunity to contribute to a book project.

Funding

Self-funding and OAPF Waiver from Intech Open Ltd.

Author contribution

RL conceived the idea. JTD drafted the manuscript. All authors contributed to reviewing the manuscript.

Conflict of interest

The authors declare no conflict of interest.

Abbreviations

B lymphocytes	Bone-marrow-derived
Dub(s)	Deubiquitinase(s)
ERAD	Endoplasmic reticulum-associated degradation
GAN	Giant axonal neuropathy
H1–3	α-helice (1 to 3)1
JAB	Jun activation domain-binding protein 1
LC–MS/MS	Liquid Chromatography with tandem mass spectrometry
MJDs/Josephins	Five Machado-Josephin domain proteases
MPN/MOV34	Mpr1, Pad1 N-terminal domain
NFL	Neurofibrillary tangles
NMR	Nuclear magnetic resonance
OUT	Ovarian tumour proteases
PHF	Paired helical filaments
PTM	Post-translational modification
RP	Ribosomal protein
T lymphocytes	Thymus derived

UBD(s)	Ubiquitin-binding domain(s)
UBP(s)	Ubiquitin-binding protein(s)
UCH	Ubiquitin *C*-terminal hydrolases
UPS	Ubiquitin-proteasome system
USP	Ubiquitin-specific proteases
ZnF_A20	Zinc finger domain of A20
ZnF-UBP	Zinc finger ubiquitin-binding protein
$\beta\text{-}_{h1}$	β-hairpin1
$\beta\text{-}_{h2}$	β-hairpin2
β-s	β-strand
$\beta_{\text{-t}}$	β-turn

Author details

Julius T. Dongdem[1*], Simon P. Dawson[2] and Robert Layfield[2]

1 Department of Chemical and Molecular Pathology, School of Medicine, University for Development Studies, Tamale, Ghana

2 School of Life Sciences, University of Nottingham Medical School, Queen's Medical Centre, Nottingham, UK

*Address all correspondence to: julius.dongdem@uds.edu.gh

References

[1] Goldstein G, Scheid M, Hammerling U, Schlesinger DH, Niall HD, Boyse EA. Isolation of a polypeptide that has lymphocyte-differentiating properties and is probably represented universally in living cells. Proceedings of the National Academy of Sciences USA. 1975;**72**(1):11-15. DOI: 10.1073/pnas.72.1.11

[2] Vijay-Kumar S, Bugg CE, Wilkinson KD, Cook WJ. Three-dimensional structure of ubiquitin at 2.8 a resolution. Proceedings of the National Academy of Sciences USA. 1985;**82**(11):3582-3585. DOI: 10.1073/pnas.82.11.3582

[3] Low TLK, Goldstein AL. The chemistry and biology of thymosin: II. Amino acid sequence analysis of thymosin α1 and polypeptide β1. The Journal of Biological Chemistry. 1979;**254**:987-995. DOI: 10.1016/S0021-9258(17)37901-2

[4] Goldknopf IL, Busch H. Isopeptide linkage between nonhistone and histone 2A polypeptides of chromosomal conjugate-protein A24. Proceedings of the National Academy of Sciences USA. 1977;**74**:864-868. DOI: 10.1073/pnas.74.3.864

[5] Ciechanover A, Gonen H, Elias S, Mayer A. Degradation of proteins by the ubiquitin-mediated proteolytic pathway. The New Biologist. 1990;**2**:227-234

[6] Ciechanover A, Heller H, Elias S, Haas AL, Hershko A. ATP-dependent conjugation of reticulocyte proteins with the polypeptide required for protein degradation. Proceedings of the National Academy of Sciences USA. 1980;**77**(3):1365-1368. DOI: 10.1073/pnas.77.3.1365

[7] Hershko A, Ciechanover A, Heller H, Haas AI, Rose IA. Proposed role of ATP in protein breakdown: Conjugation of proteins with multiple chains of the polypeptide of ATP-dependent proteolysis. Proceedings of the National Academy of Sciences USA. 1980;**77**(4):1783-1786. DOI: 10.1073/pnas.77.4.1783

[8] Hershko A, Ciechanover A. The ubiquitin system for protein degradation. Annual Review of Biochemistry. 1992;**61**:761-807. DOI: 10.1146/annurev.bi.61.070192.003553

[9] Smallridge R. The 'kiss of death' prize. Nature Reviews. Molecular Cell Biology. 2004;**5**:870. DOI: 10.1038/nrm1535

[10] Etlinger JD, Goldberg AL. A soluble ATP-dependent proteolytic system responsible for the degradation of abnormal proteins in reticulocytes. Proceedings of the National Academy of Sciences of the United States of America. 1977;**74**(1):54-58. DOI: 10.1073/pnas.74.1.54

[11] Birsa N, Norkett R, Wauer T, Mevissen TET, Wu H-C, Foltynie T, et al. Lysine 27 ubiquitination of the mitochondrial transport protein Miro is dependent on serine 65 of the parkin ubiquitin ligase. The Journal of Biological Chemistry. 2014;**289**(21):14569-14582. DOI: 10.1074/jbc.M114.563031

[12] Gegg ME, Cooper JM, Chau K-Y, Rojo M, Anthony HV, Schapira AHV, Taanman J-W. Mitofusin 1 and mitofusin 2 are ubiquitinated in a PINK1/parkin-dependent manner upon induction of mitophagy. Human Molecular Genetics. 2010;**19**(24):4861-4870. DOI: 10.1093/hmg/ddq419

[13] Dongdem JT, Wezena CA. Functional significance of the E3 ubiquitin ligases in disease and therapeutics. In: Haider S, Haider APA, Catala A, editors. Hydrolases. London, UK: IntechOpen; 2022. DOI: 10.5772/intechopen.100534

[14] Kirkin V, Dikic I. Ubiquitin networks in cancer. Current Opinion in Genetics & Development. 2011;**21**:21-28. DOI: 10.1016/j.gde.2010.10.004

[15] Tai HC, Schuman EM. Ubiquitin, the proteasome and protein degradation in neuronal function and dysfunction. Nature Reviews. Neuroscience. 2008;**9**:826-838. DOI: 10.1038/nrn2499

[16] Kane LA, Lazarou M, Fogel AI, Li Y, Yamano K, Sarraf SA, et al. PINK1 phosphorylates ubiquitin to activate parkin E3 ubiquitin ligase activity. The Journal of Cell Biology. 2014;**205**:143-153. DOI: 10.1083/jcb.201402104

[17] Kazlauskaite A, Kondapalli C, Gourlay R, Campbell DG, Ritorto MS, Hofmann K, et al. Parkin is activated by PINK1-dependent phosphorylation of ubiquitin at Ser65. The Biochemical Journal. 2014;**460**:127-139. DOI: 10.1042/BJ20140334

[18] Koyano F, Okatsu K, Kosako H, Tamura Y, Go E, Kimura M, et al. Ubiquitin is phosphorylated by PINK1 to activate parkin. Nature. 2014;**510**(7503):162-166. DOI: 10.1038/NATURE13392

[19] Lenkinski RE, Chen DM, Glickson JD, Goldstein G. Nuclear magnetic resonance studies of the denaturation of ubiquitin. Biochimica et Biophysica Acta. 1977;**494**:126-130. DOI: 10.1016/0005-2795(77)90140-4

[20] Wintrode PL, Makhatadze GI, Privalov PL. Thermodynamics of ubiquitin unfolding. Proteins. 1994;**18**:246-253. DOI: 10.1002/prot.340180305

[21] Briggs MS, Roder H. Early hydrogen-bonding events in the folding reaction of ubiquitin. Proceedings of the National Academy of Sciences USA. 1992;**89**:2017-2021. DOI: 10.1073/pnas.89.6.2017

[22] Woolfson DN, Cooper A, Harding MM, Williams DH, Evans PA. Protein folding in the absence of the solvent ordering contribution to the hydrophobic interaction. Journal of Molecular Biology. 1993;**229**:502-511. DOI: 10.1006/jmbi.1993.1049

[23] Wilkinson KD, Urban MK, Haas AL. Ubiquitin is the ATP-dependent proteolysis factor I of rabbit reticulocytes. The Journal of Biological Chemistry. 1980;**255**:7529-7532

[24] Kimura Y, Tanaka K. Regulatory mechanisms involved in the control of ubiquitin homeostasis. Journal of Biochemistry. 2010;**147**(6):793-798. DOI: 10.1093/jb/mvq044

[25] Webb GC, Baker RT, Fagan K, Board PG. Localization of the human UbB polyubiquitin gene to chromosome band 17p11.1-17p12. American Journal of Human Genetics. 1990;**46**:308-315

[26] Webb GC, Baker RT, Coggan M, Board PG. Localization of the human UBA52 ubiquitin fusion gene to chromosome band 19p13.1-p12. Genomics. 1994;**19**:567-569. DOI: 10.1006/geno.1994.1108

[27] Board PG, Coggan M, Baker RT, Vuust J, Webb GC. Localization of the human UBC polyubiquitin gene to chromosome band 12q24.3. Genomics. 1992;**12**:639-642. DOI: 10.1016/0888-7543(92)90287-3

[28] Wiborg O, Pedersen MS, Wind A, Berglund LE, Marcker KA, Vuust J. The human ubiquitin multigene family: Some genes contain multiple directly repeated ubiquitin coding sequences. The EMBO Journal. 1985;**4**:755-759. DOI: 10.1002/j.1460-2075.1985.tb03693.x

[29] Grou CP, Pinto MP, Mendes AV, Domingues P, Azevedo JE. The de novo synthesis of ubiquitin: Identification of deubiquitinases acting on ubiquitin precursors. Scientific Reports. 2015;**5**:12836. DOI: 10.1038/srep12836

[30] Dubois M-L, Meller A, Samandi S, Brunelle M, Frion J, Brunet MA, Toupin A, Beaudoin MC, Jacques J-F, Lévesque D, Scott MS, Lavigne P, Roucou X, Boisvert F-M. 1 UBB pseudogene 4 encodes functional ubiquitin variants. Nature Communications. 2020;**11**:1306. DOI: 10.1038/s41467-020-15090-6

[31] Kobayashi M, Oshima S, Maeyashiki C, Nibe Y, Otsubo K, Matsuzawa Y, et al. The ubiquitin hybrid gene UBA52 regulates ubiquitination of ribosome and sustains embryonic development. Scientific Reports. 2016;**10**(6):36780. DOI: 10.1038/srep36780

[32] Khayer N, Mirzaie M, Marashi S-A, Jalessi M. Rps27a might act as a controller of microglia activation in triggering neurodegenerative diseases. PLoS One. 2020;**15**(9):e0239219. DOI: 10.1371/journal.pone.0239219

[33] Nosrati N, Kapoor NR, Kumar V. DNA damage stress induces the expression of ribosomal protein S27a gene in a p53-dependent manner. Gene. 2015;**559**:44-51. DOI: 10.1016/j.gene.2015.01.014

[34] Houcai W, Jing Y, Lixia Z, Yuanyuan X, Shuying C, Haiyan X, et al. RPS27a promotes proliferation, regulates cell cycle progression and inhibits apoptosis of leukemia cells. Biochemical and Biophysical Research Communications. 2014;**4**:1204-1210. DOI: 10.1016/j.bbrc.2014.03.086

[35] Han S-W, Jung B-K, Ryu K-Y. Regulation of polyubiquitin genes to meet cellular ubiquitin requirement. BMB Reports. 2021;**54**(4):189-195. DOI: 10.5483/BMBRep.2021.54.4.005

[36] Finch JS, John ST, Krieg P, Bonham K, Smith HT, Fried VA, et al. Overexpression of three ubiquitin genes in mouse epidermal tumors is associated with enhanced cellular proliferation and stress. Cell Growth & Differentiation. 1992;**3**:269-278

[37] Fornace AJ, Alamo I, Hollander MC, Lamoreaux E. Ubiquitin mRNA is a major stress induced transcript in mammalian cells. Nucleic Acids Research. 1989;**17**:1215-1230. DOI: 10.1093/nar/17.3.1215

[38] Radici L, Bianchi M, Crinelli C, Magnani M. Ubiquitin C gene: Structure, function, and transcriptional regulation. Advances in Bioscience and Biotechnology. 2013;**4**:1057-1062. DOI: 10.4236/abb.2013.412141

[39] Vijay-Kumar S, Bugg CE, Cook WJ. Structure of ubiquitin refined at 1.8 a resolution. Journal of Molecular Biology. 1987;**194**(3):531-544. DOI: 10.1016/0022-2836(87)90679-6

[40] Gavilanes JG, Gonzalez de Buitrago G, Perez-Castells R, Rodriguez R. Isolation, characterization, and amino acid sequence of a ubiquitin-like protein from insect eggs. The Journal of Biological Chemistry. 1982;**257**(17):10267-10270

[41] Schlesinger DH, Goldstein G, Niall HD. Complete amino acid sequence of ubiquitin, an adenylate cyclase

stimulating polypeptide probably universal in living cells. Biochemistry. 1975;**14**(10):2214-2218. DOI: 10.1021/bi00681a026

[42] Schlesinger DH, Goldstein G. Molecular conservation of 74 amino acid sequence of ubiquitin between cattle and man. Nature. 1975;**255**:423-424. DOI: 10.1038/255423a0

[43] Watson DC, Levy WB, Dixon GH. Free ubiquitin is a non-histone protein of trout testis chromatin. Nature. 1978;**276**:196-198. DOI: 10.1038/276196a0

[44] Zuin A, Isasa M, Crosas B. Ubiquitin Signaling: Extreme conservation as a source of diversity. Cell. 2014;**3**:690-701. DOI: 10.3390/cells3030690

[45] Pickart CM. Ubiquitin enters the new millennium. Molecular Cell. 2001;**8**:499-504. DOI: 10.1016/s1097-2765(01)00347-1

[46] Jaremko L and Jaremko M. Ubiquitin. Retrieved 6th of Nov, 2022 from worldwide web http://www.chm.bris.ac.uk/motm/ubiquitin/ubiquitin_pliki/page0002.htm

[47] Trempe J-F. Reading the ubiquitin postal code. Current Opinion in Structural Biology. 2011;**21**:792-801. DOI: 10.1016/j.sbi.2011.09.009

[48] Zhan Z, Ding Y, Zhao R, Zhang Y, Yu H, Zhou Q, et al. Rapid functional divergence of a newly evolved Polyubiquitin gene in drosophila and its role in the trade-off between male fecundity and lifespan. Molecular Biology and Evolution. 2012;**29**(5):1407-1416. DOI: 10.1093/molbev/msr299

[49] Ibarra-Molero B, Makhatadze GI, Sanchez-Ruiz JM. Cold denaturation of ubiquitin. Biochimica et Biophysica Acta. 1999;**1429**(2):384-390. DOI: 10.1016/S0167-4838(98)00252-0

[50] Stockman BJ, Euvrard A, Scahill TA. Heteronuclear three-dimensional NMR spectroscopy of a partially denatured protein: The A-state of human ubiquitin. Journal of Biomolecular NMR. 1993;**3**(3):285-296. DOI: 10.1007/BF00212515

[51] Dikic I, Wakatsuki S, Walters KJ. Ubiquitin-binding domains- from structures to functions. Nature Reviews. Molecular Cell Biology. 2009;**10**:659-671. DOI: 10.1038/nrm2767

[52] Kang RS, Daniels CM, Francis SA, Shih SC, Salerno WJ, Hicke L, et al. Solution structure of a CUE-monoubiquitin complex reveals a conserved mode of ubiquitin binding. Cell. 2003;**113**:621-630. DOI: 10.1016/s0092-8674(03)00362-3

[53] Mueller TD, Kamionka M, Feigon J. Specificity of the interaction between ubiquitin-associated domains and ubiquitin. The Journal of Biological Chemistry. 2004;**279**:11926-11936. DOI: 10.1074/jbc.M312865200

[54] Laplantine E, Fontan E, Chiaravalli J, Lopez T, Lakisic G, Véron M, et al. NEMO specifically recognizes K63-linked poly-ubiquitin chains through a new bipartite ubiquitin-binding domain. The EMBO Journal. 2009;**28**:2885-2895. DOI: 10.1038/emboj.2009.241

[55] Komander D, Rape M. The ubiquitin code. Ann Rev. The Biochemist. 2012;**81**:203-229. DOI: 10.1146/annurev-biochem-060310-170328

[56] Reyes-Turcu FE, Horton JR, Mullally JE, Heroux A, Cheng X, Wilkinson KD. The ubiquitin-binding domain ZnF UBP recognizes

the C-terminal diglycine motif of unanchored ubiquitin. Cell. 2006;**124**:1197-1208. DOI: 10.1016/j.cell.2006.02.038

[57] Lee S, Tsai YC, Mattera R, Smith WJ, Michael S, Weissman AM, Bonifacino JS, Hurley JH. Structural basis for ubiquitin recognition and autoubiquitination by Rabex-5. Nature Structural & Molecular Biology. 2006;**13**:264-271. DOI: 10.1038/nsmb1064

[58] Penego L, Mapelli M, Murachelli AG, Confalonieri S, Magri L, Musacchio A, et al. Crystal structure of the ubiquitin binding domains of Rabex-5 reveals two modes of interaction with ubiquitin. Cell. 2006;**125**:1183-1195. DOI: 10.1016/j.cell.2006.02.020

[59] Crespo MD, Rubini M. Rational Design of Protein Stability: Effect of (2S,4R)-4-Fluoroproline on the stability and folding pathway of ubiquitin. PLoS One. 2011;**6**(5):e19425. DOI: 10.1371/journal.pone.0019425

[60] Amerik AY, Swaminathan S, Krantz BA, Wilkinson KD, Hochstrasser M. In vivo disassembly of free polyubiquitin chains by yeast Ubp14 modulates rates of protein degradation by the proteasome. The EMBO Journal. 1997;**16**:4826-4838. DOI: 10.1093/emboj/16.16.4826

[61] Braten O, Shabek N, Kravtsova-Ivantsiv Y, Ciechanover A. Generation of free ubiquitin chains is up-regulated in stress and facilitated by the HECT domain ubiquitin ligases UFD4 and HUL5. The Biochemical Journal. 2012;**444**:611-617. DOI: 10.1042/BJ20111840

[62] Ouyang H, Ali YO, Ravichandran M, Dong A, Qiu W, MacKenzie F, et al. Protein aggregates are recruited to aggresome by histone deacetylase 6 via unanchored ubiquitin C termini. The Journal of Biological Chemistry. 2012;**287**:2317-2327. DOI: 10.1074/jbc.M111.273730

[63] Cook WJ, Jeffrey LC, Carson M, Chen Z, Pickart CM. Structure of a diubiquitin conjugate and a model for interaction with ubiquitin conjugating enzyme (E2). The Journal of Biological Chemistry. 1992;**267**:16467-16471. DOI: 10.2210/pdb1aar/pdb

[64] Pickart CM, Fushman D. Polyubiquitin chains: Polymeric protein signals. Current Opinion in Chemical Biology. 2004;**8**(6):610-616. DOI: 10.1016/j.cbpa.2004.09.009

[65] Johnson ES. Ubiquitin branches out. Nature Cell Biology. 2002;**4**:295-298. DOI: 10.1038/ncb1202-e295

[66] Haas AL, Ahrens P, Bright PM, Ankel H. Interferon induces a 15-kilodalton protein exhibiting marked homology to ubiquitin. The Journal of Biological Chemistry. 1987;**262**(23):11315-11323

[67] Davis DL, Soldin SJ. Identification of ubiquitin as an immunophilin. Biochemical and Biophysical Research Communications. 2000;**277**(2):325-329. DOI: 10.1006/bbrc.2000.3684

[68] Szewczuk Z, Stefanowicz P, Wilczynski A, Staszewska A, Siemion IZ, Zimecki M, et al. Immunosuppressory activity of ubiquitin fragments containing retro-RGD sequence. Biopolymers. 2004;**74**(5):352-362. DOI: 10.1002/bip.20084

[69] Manetto V, Abdul-Karim FW, Perry G, Tabaton M, Autilio-Gambetti L, Gambetti P. Selective presence of ubiquitin in intracellular inclusions. The American Journal of Pathology. 1989;**134**(3):505-513

Cell Biology. 1995;**7**(2):215-223. DOI: 10.1016/0955-0674(95)80031-x

[102] Clague MJ, Barsukov I, Coulson JM, Liu H, Rigden DJ, Urbé S. Deubiquitylases from genes to organism. Physiological Reviews. 2013;**93**:1289-1315. DOI: 10.1152/physrev.00002.2013

[103] Wilkinson KD. Regulation of ubiquitin-dependent processes by deubiquitinating enzymes. The FASEB Journal. 1997;**11**:1245-1256. DOI: 10.1096/fasebj.11.14.9409543

[104] Baker RT, Board PG. The human ubiquitin gene family: Structure of a gene and pseudogenes from the Ub B subfamily. Nucleic Acids Research. 1987;**15**:443-463. DOI: 10.1093/nar/15.2.443

[105] Ozkaynak E, Finley D, Solomon MJ, Varshavsky A. The yeast ubiquitin genes: A family of natural gene fusions. The EMBO Journal. 1987;**6**:1429-1439. DOI: 10.1002/j.1460-2075.1987.tb02384.x

[106] Finley D, Bartel B, Varshavsky A. The tails of ubiquitin precursors are ribosomal proteins whose fusion to ubiquitin facilitates ribosome biogenesis. Nature. 1989;**338**:394-401. DOI: 10.1038/338394a0

[107] Tobias JW, Varshavsky A. Cloning and functional analysis of the ubiquitin-specific protease gene UBP1 of Saccharomyces cerevisiae. The Journal of Biological Chemistry. 1991;**266**:12021-12028. DOI: 10.1016/S0021-9258(18)99059-9

[108] Piotrowski J, Beal R, Hoffman L, Wilkinson KD, Cohen RE, Pickart CM. Inhibition of the 26S proteasome by polyubiquitin chains synthesized to have defined lengths. The Journal of Biological Chemistry. 1997;**272**:23712-23721. DOI: 10.1074/jbc.272.38.23712

[109] Pickart CM, Rose IA. Ubiquitin carboxyl-terminal hydrolase acts on ubiquitin carboxyl-terminal amides. The Journal of Biological Chemistry. 1985;**260**:7903-7910. DOI: 10.1016/S0021-9258(17)39538-8

[110] Rahighi S, Dikic I. Selectivity of the ubiquitin-binding modules. FEBS Letters. 2012;**586**:2705-2710. DOI: 10.1016/j.febslet.2012.04.053

[111] Huguenin-Dezot N, De Cesare V, Peltier J, Kulathu Y, Trost M, Chin JW. Synthesis of isomeric phosphoubiquitin chains reveals that phosphorylation controls deubiquitinase activity and specificity. Cell Reports. 2016;**16**:1180-1193. DOI: 10.1016/j.celrep.2016.06.064

[112] Lee S, Tumolo JM, Ehlinger AC, Jernigan KK, Qualls-Histed SJ, Hsu PC, et al. Ubiquitin turnover and endocytic trafficking in yeast are regulated by Ser57 phosphorylation of ubiquitin. eLife. 2017;**13**:6. pii: e29176. DOI: 10.7554/eLife.29176

[113] Schaefer JB, Morgan DO. Protein-linked ubiquitin chain structure restricts activity of deubiquitinating enzymes. The Journal of Biological Chemistry. 2011;**286**(52):45186-45196. DOI: 10.1074/jbc.M111.310094

[114] Swaney DL, Rodríguez-Mias RA, Villén J. Phosphorylation of ubiquitin at Ser65 affects its polymerization, targets, and proteome-wide turnover. EMBO Reports. 2015;**16**:1131-1144. DOI: 10.15252/embr.201540298

[115] Wauer T, Swatek KN, Wagstaff JL, Gladkova C, Pruneda JN, Michel MA, et al. Ubiquitin Ser65 phosphorylation affects ubiquitin structure, chain assembly and hydrolysis. The EMBO Journal. 2015;**34**:307-325. DOI: 10.15252/embj.201489847

[116] Hoeller D, Dikic I. Regulation of ubiquitin receptors by coupled monoubiquitination. Sub-Cellular Biochemistry. 2010;**54**:31-40. DOI: 10.1007/978-1-4419-6676-6_3

[117] Varadan R, Assfalg M, Raasi S, Pickart C, Fushman D. Structural determinants for selective recognition of a Lys48-linked polyubiquitin chain by a UBA domain. Molecular Cell. 2005;**18**:687-698. DOI: 10.1016/j.molcel.2005.05.013

[118] Andersen KM, Hofmann K, Hartmann-Petersen R. Ubiquitin-binding proteins: Similar, but different. Essays in Biochemistry. 2005;**41**:49-67. DOI: 10.1042/EB0410049

[119] Hicke L, Schubert HL, Hill CP. Ubiquitin-binding domains. Nature Reviews. Molecular Cell Biology. 2005;**6**:610-621. DOI: 10.1038/nrm1701

[120] Haglund K, Stenmark H. Working out coupled monoubiquitination. Nature Cell Biology. 2006;**8**:1218-1219. DOI: 10.1038/ncb1106-1218

[121] Deveraux Q, Ustrell V, Pickart C, Rechsteiner M. A 26S protease subunit that binds ubiquitin conjugates. The Journal of Biological Chemistry. 1994;**269**:7059-7061

[122] Lam YA, Lawson TG, Velayutham M, Zweier JL, Pickart CM. A proteasomal ATPase subunit recognizes the polyubiquitin degradation signal. Nature. 2002;**416**:763-767. DOI: 10.1038/416763a

[123] Young P, Deveraux Q, Beal RE, Pickart CM, Rechsteiner M. Characterization of two polyubiquitin binding sites in the 26S protease subunit 5a. The Journal of Biological Chemistry. 1998;**273**:5461-5467. DOI: 10.1074/jbc.273.10.5461

[124] Schreiner P, Chen X, Husnjak K, Randles L, Zhang N, Elsasser S, et al. Ubiquitin docking at the proteasome through a novel pleckstrin-homology domain interaction. Nature. 2008;**453**:548-552. DOI: 10.1038/nature06924

[125] Kirkin V, Lamark T, Sou YS, Bjørkøy G, Nunn JL, Bruun JA, et al. A role for NBR1 in autophagosomal degradation of ubiquitinated substrates. Molecular Cell. 2009;**33**:505-516. DOI: 10.1016/j.molcel.2009.01.020

[126] Rahighi S, Ikeda F, Kawasaki M, Akutsu M, Suzuki N, Kato R, et al. Specific recognition of linear ubiquitin chains by NEMO is important for NF-κB activation. Cell. 2009;**136**:1098-1109. DOI: 10.1016/j.cell.2009.03.007

[127] Wagner S, Carpentier I, Rogov V, Kreike M, Ikeda F, Löhr F, et al. Ubiquitin binding mediates the NF-κB inhibitory potential of ABIN proteins. Oncogene. 2008;**27**:3739-3745. DOI: 10.1038/sj.onc.1211042

[128] Bellare P, Small EC, Huang X, Wohlschlegel JA, Staley JP, Sontheimer EJ. A role for ubiquitin in the spliceosome assembly pathway. Nature Structural & Molecular Biology. 2008;**15**:444-451. DOI: 10.1038/nsmb.1401

[129] Brzovic PS, Lissounov A, Christensen DE, Hoyt DW, Klevit RE. A UbcH5/ubiquitin noncovalent complex is required for processive BRCA1-directed ubiquitination. Molecular Cell. 2006;**21**(6):873-880. DOI: 10.1016/j.molcel.2006.02.008

[130] Crespo MD, Platt GW, Bofill R, Searle MS. Context-dependent effects of proline residues on the stability and folding pathway of ubiquitin. European Journal of

Biochemistry. 2004;**271**:4474-4484. DOI: 10.1111/j.1432-1033.2004.04392.x

[131] Husnjak K, Dikic I. Ubiquitin binding proteins: Decoders of ubiquitin-mediated cellular functions. Annual Review of Biochemistry. 2012;**81**:291-322. DOI: 10.1146/annurev-biochem-051810-094654

[132] Herhaus L, Dikic I. Expanding the ubiquitin code through post-translational modification. EMBO Reports. 2015;**16**:1071-1083. DOI: 10.15252/embr.201540891

[133] Ohtake F, Saeki Y, Sakamoto K, Ohtake K, Nishikawa H, Tsuchiya H, et al. Ubiquitin acetylation inhibits polyubiquitin chain elongation. EMBO Reports. 2015;**16**:192-201. DOI: 10.15252/embr.201439152

[134] Sharma K, D'Souza RCJ, Tyanova S, Schaab C, Wiśniewski JR, Cox J, et al. Ultradeep human Phosphoproteome reveals a distinct regulatory nature of Tyr and Ser/Thr-based Signaling. Cell Reports. 2014;**8**(5):1583-1594. DOI: 10.1016/j.celrep.2014.07.036

[135] Bian Y, Song C, Cheng K, Dong M, Wang F, Huang J, et al. An enzyme assisted RP-RPLC approach forin-depth analysis of human liver phosphoproteome. Journal of Proteomics. 2014;**96**:253-262. DOI: 10.1016/j.jprot.2013.11.014

[136] Gu TL, Deng X, Huang F, Tucker M, Crosby K, Rimkunas V, et al. Survey of tyrosine kinase signaling reveals ROS kinase fusions in human cholangiocarcinoma. PLoS One. 2011;**6**(1):e15640. DOI: 10.1371/journal.pone.0015640

[137] Pan C, Olsen JV, Daub H, Mann M. Global effects of kinase inhibitors on signaling networks revealed by quantitative phosphoproteomics. Molecular & Cellular Proteomics. 2009;**8**:2796-2808. DOI: 10.1074/mcp.M900285-MCP200

[138] Shiromizu T, Adachi J, Watanabe S, Murakami T, Kuga T, Muraoka S, et al. Identification of missing proteins in the nextprot database and unregistered phosphopeptides in the phosphositeplus databas as part of the chromosome-centric human proteome project. Journal of Proteome Research. 2013;**12**:2414-2421. DOI: 10.1021/pr300825ve

[139] Swaney DL, Beltrao P, Starita L, Guo A, Rush J, Fields S, et al. Global analysis of phosphorylation and ubiquitylation cross-talk in protein degradation. Nature Methods. 2013;**10**:676-682. DOI: 10.1038/nmeth.2519

[140] Zhou H, Di Palma S, Preisinger C, Peng M, Polat AN, Heck AJ, et al. Toward a comprehensive characterization of a human cancer cell phosphoproteome. Journal of Proteome Research. 2013;**12**:260-271. DOI: 10.1021/pr300630k

Chapter 2

Ubiquitination on the Peroxisomal Membrane for Protein Transport in Plants

Shoji Mano, Kazumi Hikino and Masatake Kanai

Abstract

Peroxisomes are ubiquitous organelles present in most eukaryotic cells that have important biological functions related to fatty acid metabolism and detoxification of reactive oxygen species. Disruption of peroxisomal function affects the survival of cells and organisms. Peroxisomes do not have their own genome, and peroxisomal proteins are encoded in the nuclear genome. Therefore, efficient and accurate post-translational transport of peroxisomal proteins is necessary to maintain peroxisomal function. In mammals, yeast, and plants, many factors involved in protein transport to peroxisomes have been identified and their molecular mechanisms elucidated. In plants, analysis of Arabidopsis peroxisome mutants, such as *apem* (*aberrant peroxisome morphology*) and *ibr* (*indole-3-butyric acid-response*), enabled the identification of the factors mediating protein transport. Of these, several proteins, such as PEX1 (Peroxin 1), PEX2, PEX4, PEX6, PEX10, PEX12, PEX22, and APEM9, constitute the ubiquitin system on the peroxisomal membrane, and loss of function of each protein reduces the efficiency of protein transport to peroxisomes. This ubiquitin-dependent peroxisomal protein transport system is also present in yeast and mammalian cells and is an example of a type of ubiquitin modification that serves as a signaling tag rather than as a tag for protein degradation. This chapter introduces the factors involved in protein transport to the peroxisome via the ubiquitin system in plants and outlines their functions.

Keywords: *Arabidopsis thaliana*, peroxisome, peroxin, protein transport, ubiquitin system

1. Introduction

Peroxisomes are ubiquitous organelles present in most eukaryotic cells; they were discovered in the early 1960s as organelles measuring approximately 1.0 μm in diameter [1]. Peroxisomes proliferate by division of preexisting peroxisomes, and abnormal peroxisomes that are no longer needed or have been oxidized are degraded via a peroxisome-specific autophagy process called pexophagy [2–7]. Peroxisomes are involved in various biological functions. Of these, fatty acid metabolism and detoxification of reactive oxygen species are functions common to many organisms, whereas bile acid biosynthesis and alcohol metabolism are peroxisome functions specific to mammals and yeast, respectively. Photorespiration, which salvages byproducts of photosynthesis, and biosynthesis of phytohormones such as jasmonic acid and auxin are plant peroxisome-specific functions. Peroxisomes do not have their own genome and all peroxisomal proteins are encoded in the nuclear genome; therefore, post-translational transport must be efficient and accurate to ensure peroxisomal protein function. Numerous studies in a variety of organisms have identified the processes involved in peroxisome biogenesis, such as protein transport, proliferation, differentiation, and inheritance, and have elucidated the molecular mechanisms [8–12]. In particular, the analysis of peroxisome biogenesis factors called peroxins (PEXs) revealed the molecular mechanism underlying peroxisome biogenesis including protein transport [8–12].

Peroxisomal protein transport can be divided into the following stages (**Figures 1** and **2**). (i) Cytosolic receptors recognize and bind peroxisomal proteins,

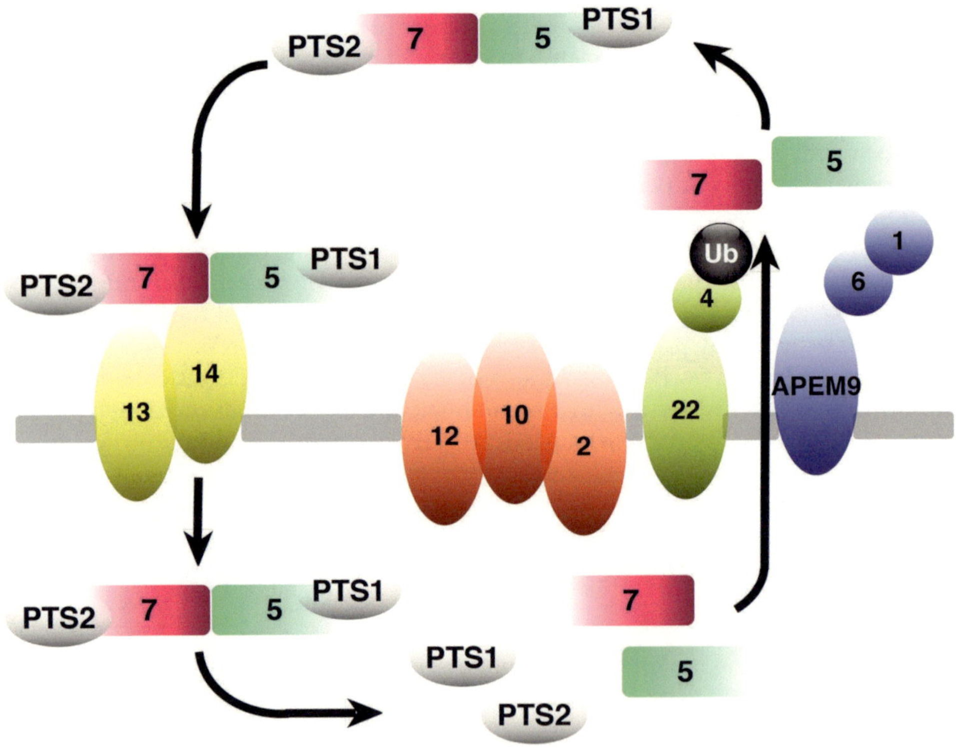

Figure 1.
Protein transport to peroxisomes in plants. Most peroxisomal proteins contain PTS1 or PTS2. PTS1 and PTS2 are recognized by the cytosolic receptors PEX5 and PEX7, respectively, and are directed to a docking complex containing PEX13 and PEX14 on the peroxisomal membrane. The peroxisomal protein-receptor complex passes through the peroxisomal membrane, and the peroxisomal protein and receptor are released. PEX5 and PEX7 return to the cytosol to engage in the next round of protein transport. Ubiquitination may be required for recycling the receptor in the cytosol, and proteins targeted for ubiquitination on the peroxisomal membrane are involved in the export apparatus. PEX4 has UBC activity and functions as an E2 enzyme, and PEX2, PEX10, and PEX12 are RING-finger proteins that function as E3 ligases. PEX1 and PEX6 are AAA+ ATPases involved in the dissociation of the receptor from the membrane. APEM9 is a functional homolog of yeast Pex15p and mammalian PEX26 that tethers the PEX1-PEX6 complex to the peroxisomal membrane. PEX22 is responsible for tethering PEX4 to the peroxisomal membrane. Whether the peroxisomal protein-receptor complex fully enters the peroxisome in plants remains unclear. Ubiquitination of PEX5 has been detected in yeast and animals, but not in plants.

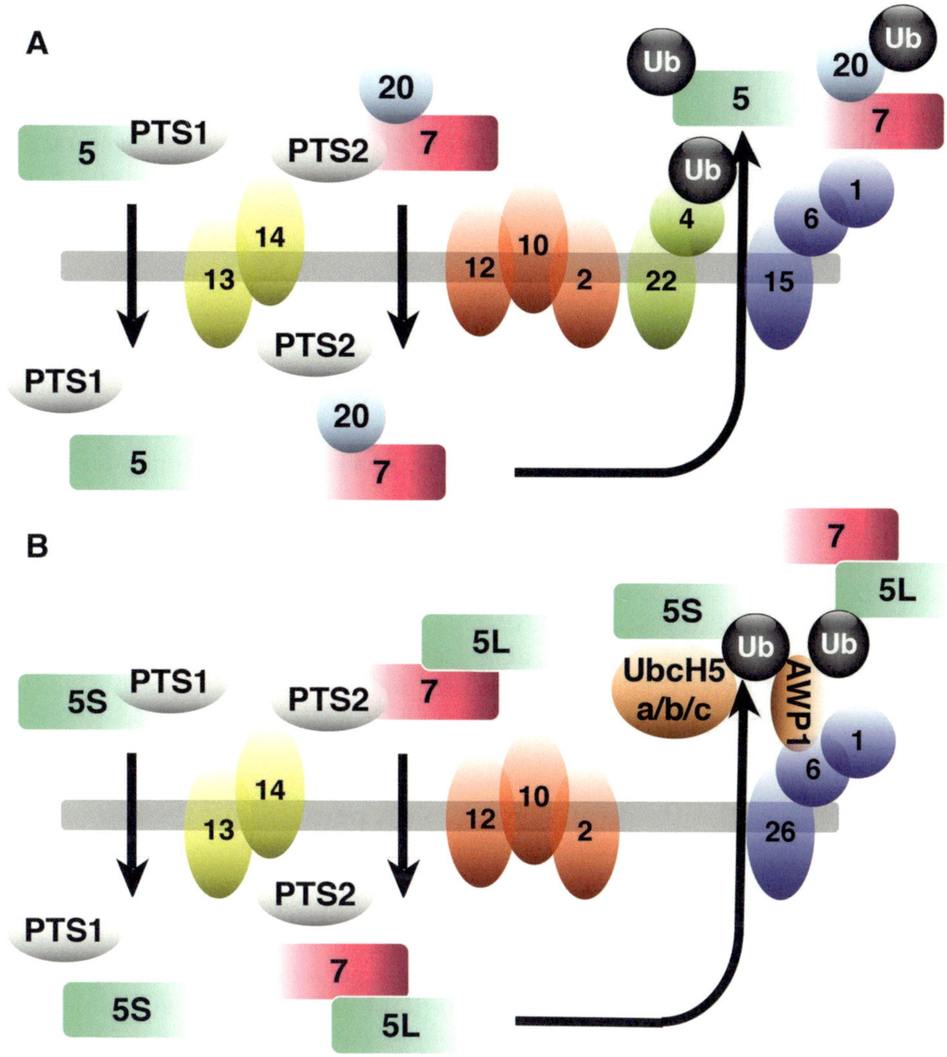

Figure 2.
Outline of PTS1- and PTS2-mediated transport in yeast and mammals. (A) In Pichia pastoris, proteins containing PTS1 and PTS2 are recognized by Pex5p and Pex7p, respectively, and they are transported independently. A co-receptor is required for efficient PTS2 transport. Pex18p and Pex21p in Saccharomyces cerevisiae and Pex20p in Pichia pastors are coreceptors of Pex7p. Pex5p and Pex20p are ubiquitinated and transported from peroxisomes to the cytosol. (B) In mammals, alternative splicing of the PEX5 gene generates two variants, PEX5S and PEX5L. PEX5S is involved in the transport of PTS1, whereas PEX5L binds to PEX7 to facilitate the efficient transport of PTS2. As in yeast, PEX5 is ubiquitinated and transported from peroxisomes to the cytosol. Proteins homologous to PEX4 are not present in mammalian genomes, and instead UbcH5a/b/c catalyzes the ubiquitination of PEX5. AWP1 functions as an adaptor for PEX6 and recognizes ubiquitinated PEX5. Pex15p, PEX26, and APEM9 in **Figure 1** *have the same function, namely, interacting with PEX6 to tether the PEX1- PEX6 complex to the peroxisomal membrane. However, the amino acid sequence identity between the three proteins is low except for the transmembrane region.*

and direct them to the docking site on the peroxisomal membrane. (ii) The peroxisomal protein-receptor complex associates with the docking complex and passes through the peroxisomal membrane. (iii) After the peroxisomal protein dissociates from the receptor, the receptor is returned to the cytosol to engage in the next transport cycle. Studies in mammals, yeast, and plants identified the factors involved in each stage: PEX5 and PEX7 in stage (i), PEX13 and PEX14 in stage (ii), and PEX1, PEX2, PEX4, PEX6, PEX10, PEX12, PEX22, and APEM9/Pex15p/PEX26 in stage (iii) (**Figures 1** and **2**) [9–12]. Various protein modification systems involving PEXs are required for efficient protein transport to peroxisomes. For example, stage (iii) requires ubiquitination for receptors, such as PEX5, on the peroxisomal membrane, and several proteins necessary for ubiquitination have been identified and characterized at the molecular level [13–16]. Defects in individual components of this ubiquitin system decrease the efficiency of protein transport to peroxisomes, leading to abnormalities such as dwarfism in plants, growth defects in yeast, and genetic diseases in humans [10, 14–16].

Ubiquitination on the peroxisomal membrane does not lead to protein degradation but has a signaling function. Ubiquitination is a posttranslational protein modification in which a ubiquitin protein of approximately 9 kDa binds to a target protein. The binding of ubiquitin to amino acids occurs through a three-step enzymatic cascade. A ubiquitin-activating enzyme (E1) activates ubiquitin via an AMP-bound intermediate. Ubiquitin then binds to the active-site cysteine in a ubiquitin-conjugating enzyme (E2) and is transferred to the target protein by a ubiquitin-protein ligase (E3). The E3 ligase can be a RING-type enzyme that binds to both the substrate and a ubiquitin-charged E2 enzyme and directly transfers the ubiquitin moiety to the target protein, or a thioester-bound HECT-type and RBR-type E3 that forms ubiquitin-E3 intermediates. The regulatory mechanism of the ubiquitination process has been described in other chapters and reviews. This chapter focuses on the ubiquitin system involved in protein transport to peroxisomes based on findings in *Arabidopsis thaliana*.

2. Components of the ubiquitin system on the peroxisomal membrane in plants

There are two types of protein transport to peroxisomes, peroxisome targeting signal (PTS) 1-dependent protein transport and PTS2-dependent protein transport. PTS1, which consists of 3–4 amino acid residues located at the C-terminal end of each protein, is recognized by the cytosolic receptor PEX5. On the other hand, PTS2-containing proteins are translated as larger precursors containing the extension sequence; PTS2 is present in the N-terminal extension sequence and is recognized by its receptor, PEX7. This extension sequence is cleaved after import into peroxisomes, and the PTS2-containing proteins become the mature form [17–19]. However, different species require different PEXs or use the same PEX in different ways. In yeast, the PTS1- and PTS2-dependent transport systems function independently from each other (**Figure 2a**). PTS2-dependent transport requires additional proteins that act as co-receptors to facilitate the import process. These include Pex18p and Pex21p in *S. cerevisiae* [20] and Pex20p in *Pichia pastoris* [21]; Pex5p and Pex20p are ubiquitinated and recycled to the cytosol [21, 22]. In mammals, two types of PEX5, PEX5L, and PEX5S are generated by alternative splicing from a single gene. PEX5L is involved in both PTS1- and PTS2-dependent protein transport, whereas PEX5S transports only PTS1 proteins (**Figure 2b**). As in yeast, ubiquitination of PEX5 is required for receptor export [23, 24], and the AWP1 (associated with PRK1) protein stimulates PEX5 export by interacting with PEX6 [25]. In plants, on the other hand, PTS1- and PTS2-dependent protein transports are interdependent; proteins containing PTS1 and PTS2 are transported as a single complex (**Figure 1**) [10, 11, 26]. Thus, although there are differences in the way PEXs are used by different organisms, they all share a common need for the ubiquitin system in protein transport. As detailed below, proteins involved in the ubiquitin system are present on the peroxisomal membrane and are required for protein transport to peroxisomes [10, 14–16]. Of these, PEX4 possesses ubiquitin-conjugating (UBC) activity [15, 16, 27–30], and PEX2, PEX10, and PEX12 function as ubiquitin ligases [31–33]. PEX1 and PEX6, which belong to the ATPases associated with diverse cellular activities (AAA+) family (AAA ATPases), are responsible for the energy required for the receptor to dissociate from the peroxisomal membrane [34, 35]. PEX22 and Pex15p/PEX26/APEM9 tether PEX4 and the PEX1-PEX6 complex to the peroxisomal membrane, respectively [36, 37].

This section outlines the function of PEX proteins involved in ubiquitination on plant peroxisomal membranes, based primarily on the analysis of Arabidopsis mutants.

2.1 Advances in plant peroxisome research achieved using Arabidopsis mutants

Analysis of Arabidopsis mutants such as *apem* and *ibr* led to dramatic advances in plant peroxisome research. The *apem* mutants were isolated from a pool of ethyl methanesulfonate (EMS)-mutagenized Arabidopsis seeds expressing the peroxisome marker *GFP-PTS1* as the parent plant. The *apem* mutants show alterations in the pattern of GFP fluorescence as follows: (i) elongated peroxisomes, (ii) enlarged peroxisomes, (iii) mislocalization of the GFP-PTS1 protein to the cytosol, and (iv) altered distribution of GFP-labeled peroxisomes [15, 38–42]. Among *apem* mutants, *apem2*, *apem4*, *apem7*, and *apem9* show GFP fluorescence not only in peroxisomes but also in the cytosol, indicating that the efficiency of PTS1-dependent protein transport to peroxisomes is reduced (**Figure 3**) [15, 39, 41]. Furthermore, PTS2-dependent protein transport is disturbed in these mutants [15, 39, 41]. Identification of *APEM* genes and analysis of their gene products revealed that *APEM2*, *APEM4*, *APEM7*, and *APEM9* encode PEX13, PEX12, PEX4, and the functional homolog of Pex15p/PEX26, respectively [15, 39, 41]. PEX4 and PEX12 function as a UBC enzyme and a ubiquitin ligase, respectively, and Pex15p/PEX26 is responsible for tethering the AAA ATPase complex, PEX1-PEX6, to the peroxisomal membrane [15, 39, 41]. APEM4/PEX12, APEM7/PEX4, and APEM9/Pex15p/PEX26 thus constitute a ubiquitin system on the peroxisomal membrane, which suggests that ubiquitination on the peroxisomal membrane is required for efficient protein transport to peroxisomes.

A screening based on peroxisome function contributed to the identification of several peroxisome-related genes encoding proteins associated with the ubiquitin system. Mutants were isolated based on the conversion of indole-3-butyric acid (IBA) to indole-3-acetic acid (IAA), an endogenous auxin, by peroxisomal fatty acid β-oxidation, an exclusive function of peroxisomes in plants. The Arabidopsis *pex1*, *pex4*, *pex6*, *pex12*, and *pex26* mutants were isolated because they exhibited abnormal conversion of IBA to IAA [30, 43–45]. A *pex1-1* mutant was isolated as a suppressor that restores metabolic and physiological defects in the *pex6* mutant [46]. Arabidopsis *PEX2* was identified as a suppressor of the *de-etiolated 1-1* (*det1-1*) mutant, which is defective in photomorphogenesis [47], and a different allele of *pex6* was identified as a *pfl* (*persistent GFP-ICL fluorescence*) mutant showing a different pattern of GFP fluorescence from the parent plant expressing the *GFP-ICL* fusion gene [48].

Many plant PEX mutants with T-DNA insertions are lethal because they show complete protein dysfunction. However, the *apem*, *ibr*, and *pfl* mutants have a milder loss of function and thus provide a tool to analyze the function of the causative gene product because these are EMS-induced nucleotide substitution mutants. Analysis of

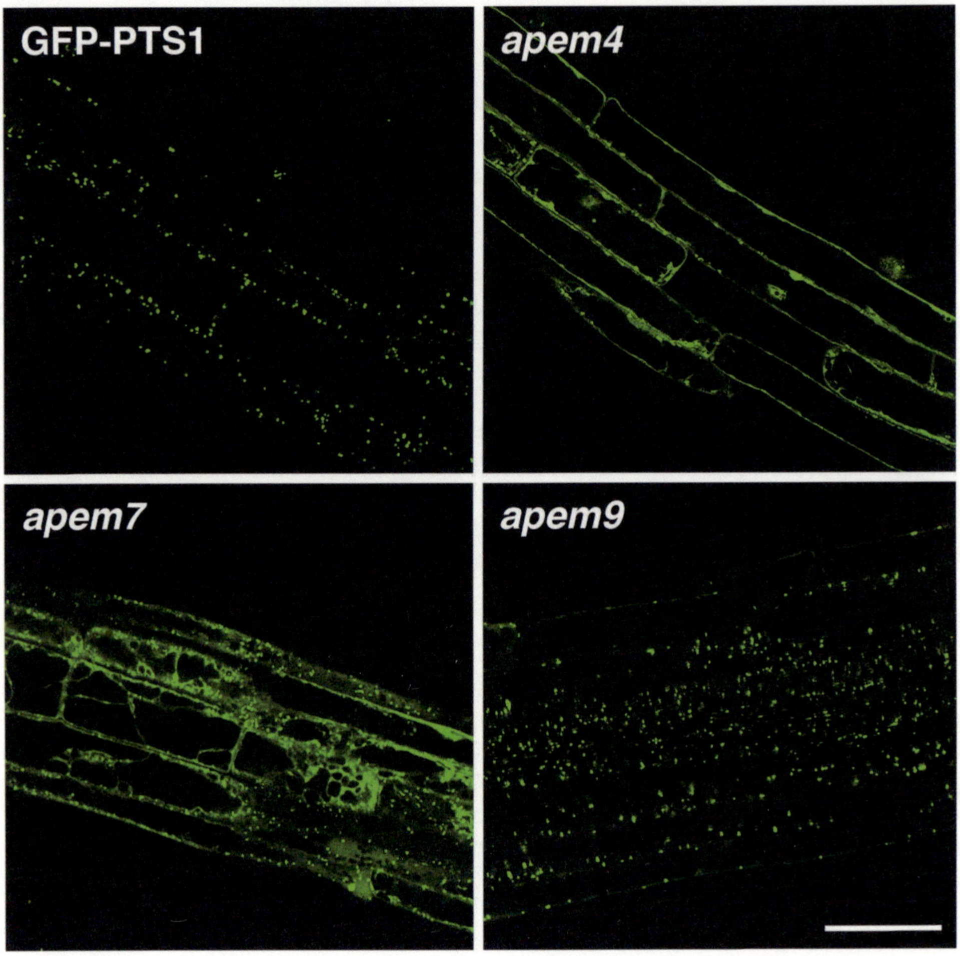

Figure 3.
GFP fluorescence patterns in root cells of the parent plant, GFP-PTS1, and apem mutants expressing the peroxisome marker GFP-PTS1. GFP is transported via a PTS1-dependent pathway, and peroxisomes are visualized as spherical structures. In the apem4, apem7, and apem9 mutants, GFP is detected in the cytosol in addition to peroxisomes. These phenotypes indicate a defect in PTS1-dependent protein transport. Bar, 50 μm.

these EMS-treated Arabidopsis mutants led to research into the role of peroxisomal membrane ubiquitin in plants. Below is an overview of the components of the ubiquitin system that regulate protein transport to peroxisomes.

2.2 PEX4 functions as a ubiquitin-conjugating enzyme

PEX4 has only been identified in yeast (also called Ubc10p in *S.cerevisiae*) and plants, and proteins with amino acid sequence homology to PEX4 have not been found in mammals. *PEX4* genes were originally identified in a screening of mutants showing growth defects when cultured under carbon-limited conditions in yeast [27–29]. In plants, the *Arabidopsis pex4* mutant was isolated separately as the *pex4-1* and *pex4-2* mutants and the *apem7* mutant [15, 30, 49]. PEX4 proteins in yeast and plants have a putative active-site cysteine residue that is essential for forming a thioester bond with ubiquitin. The UBC activity of PEX4 was identified only in yeast, where it was reported to function as an E2 enzyme in the ubiquitination process [27]. A recent analysis using rabbit reticulocyte lysate and *in vitro* ubiquitin assays with recombinant wild-type PEX4 and mutant PEX4, which has an amino acid substitution at position 123 from proline to leucine, suggests that both PEX4 forms have UBC activities, although the activity and specificity of mutant-type PEX4 are reduced [15, 16]. However, the thioester bond between mutant PEX4 and ubiquitin is cleaved *in vitro* under reducing conditions, such as the addition of β-mercaptoethanol, whereas *in vivo* cleavage is incomplete [15]. This may be due to an alteration in the interaction of PEX4 with other factors on the peroxisomal membrane and/or in the conformation of PEX4 itself caused by the *apem7/pex4* mutation. Although Arabidopsis PEX4 is difficult to purify as a soluble protein, it can be purified by expressing it as a *PEX4-PEX22* fusion gene and can then be crystallized and have its structure elucidated (PEX22 is described in the next section) [16]. Crystallographic analysis indicates that the *pex4-1* mutation site is located on a loop near the active-site cleft and it is thought to alter the residue immediately following the "gateway residue" involved in regulating ubiquitin access to the active site [16]. In other words, the proximity of the *pex4-1* mutation site to the active-site cysteine likely affects the UBC activity of mutant-type PEX4.

2.3 PEX22 functions in tethering PEX4 to the peroxisome membrane

Similar to PEX4, which has not been found in mammals, PEX22 has not been found in mammalian genomes. Analysis of the *apem7* and *pex4-1* mutants indicated that PEX4 localizes to the peroxisomal membrane [15, 30]. PEX4 does not have an obvious transmembrane domain, suggesting that it requires other factors for localization to the peroxisomal membrane. In yeast, Pex22p was identified as a factor that anchors Pex4p to the peroxisomal membrane [36, 37], and the cytosolic domain of Pex22p is involved in Pex4p activity [50]. In plants, Arabidopsis PEX22 was identified from the *Arabidopsis* cDNA library in a yeast two-hybrid analysis, which showed that it interacts with Arabidopsis PEX4 [30]. Arabidopsis PEX22 has high amino acid sequence homology with other plant homologs, but only 6–12% sequence identity with the yeast Pex22p protein; however, there are similarities in the sequence of the N-terminal transmembrane domain, the molecular weight, and the topology of the proteins [30]. *S. cerevisiae* Pex22p does not interact with *Pichia pastoris* Pex4p [36]. In addition, although coexpression of the Arabidopsis *PEX4* and *PEX22* genes complements the defects in the yeast *pex4* pex22 double mutant, PEX4 or PEX22 alone do not restore the phenotype of the corresponding yeast mutant, indicating that Arabidopsis PEX4 does not interact with yeast Pex22p and vice versa [30].

The crystal structure of Pex4p bound to the soluble portion of Pex22p has been reported in *S. cerevisiae*, and binding of Pex4p to Pex22p increases the ability of Pex4p to transfer ubiquitin to its substrates [51, 52]. In plants, the Arabidopsis PEX4-PEX22 complex was crystallized and its structure was elucidated, as described in the previous section [16]. Comparison with yeast Pex22p shows that PEX22 maintains a similar Rossmann fold structure, with several salt bridges and long unstructured tethers in positions that contribute to the specificity of PEX22 for PEX4 and the ubiquitination of peroxisomal membrane targets away from PEX4 without dissociation from PEX22 [16]. These results indicate that the interaction between PEX4 and PEX22 has been acquired with species specificity.

2.4 PEX2, PEX10, and PEX12 function as ubiquitin ligases

There are three RING-finger domain-containing ubiquitin ligases on the peroxisomal membrane, PEX2, PEX10, and PEX12, and they have been identified in various organisms including plants [9, 10, 14]. They function as E3 ubiquitin ligases during ubiquitination on the peroxisomal membrane [14, 31, 32]. In plants, Arabidopsis PEX2, PEX10, and PEX12 possess *in vitro* ubiquitin ligase activity [33]. These three RING E3 ligases depend on each other for stability, and deletion or partial loss of function of one component induces instability of the other two components [44, 53–55].

Arabidopsis *PEX2* was originally identified in an analysis of a suppressor of the *det1-1* mutant [47]. DET1 is a global repressor of photomorphogenesis, and the *det1* mutant grows in the dark with the phenotype of a light-grown plant [56]. The *pex2* mutant, originally known as *ted3* for reversal of the *det* phenotype, was identified as one of the *ted* mutants [47]. The *ted3* mutant has an amino acid substitution at position 275 from valine to methionine in the RING-finger domain that suppresses the phenotypic defect of the *det* mutant [47]. Overexpression of the PEX2 RING domain in the *det1* mutant also partially suppresses the *det1* phenotype, suggesting that peroxisomes are involved in photomorphogenesis [57]. The Arabidopsis *pex2-1* mutant was isolated by screening for abnormal degradation of the peroxisomal matrix protein [55]. Knockdown of the *PEX2* gene by RNA interference [8] showed that the *pex2-1* mutant is defective in protein transport to peroxisomes [55].

The Arabidopsis *PEX10* gene was identified as a gene encoding a protein with 47–56% sequence similarity to the product of the *PEX10* gene from mammals and yeast [58, 59]. Analysis of T-DNA insertion mutants of the *PEX10* gene showed that loss of PEX10 affects normal embryo development and viability [58, 59]. Furthermore, the *pex10-2* mutant was isolated as one of the *ibr* mutants, and further analysis showed that the *pex2-1 pex10-2* double mutant has more severe growth defects than the respective single mutants, indicating that PEX2 and PEX10 function in a coordinated manner [55]. The *pex10* mutant, which was generated by RNAi interference, shows defects in ER morphology and cuticular wax accumulation [60].

For PEX12, a T-DNA knockout mutant of PEX12 and a *pex12* partial loss-of-function mutant were generated using RNA interference, and the phenotypes of individual plants were analyzed [61]. These mutants show developmental arrest, plant growth inhibition, and reduced lethality in early embryogenesis [61]. In addition, the *apem4* (previously known as *apm4*) mutant, which shows a decrease of both PTS1- and PTS2-dependent protein transport to peroxisomes, was isolated as having a mutation in the *PEX12* gene [39]. The *apem4* and other Arabidopsis *pex12* mutants show defects in PTS1- and PTS2-dependent protein transport [39, 44], suggesting that the ubiquitin ligase activity of PEX12 is required for efficient protein transport to peroxisomes. Indeed, Arabidopsis PEX12 can bind to PEX7 [62], suggesting that PEX12 is involved in receptor recycling from peroxisomes to the cytosol, as discussed below. The *apem4* and *pex12-1* mutants are characterized by the replacement of the arginine at position 170 by lysine and that of glutamic acid at position 171 by lysine, respectively [39, 44]. The mechanism by which these mutations affect the ligase activity and conformation of PEX12 remains to be elucidated.

2.5 PEX1 and PEX6 function as AAA ATPases

PEX1 and PEX6 are type 2 AAA+ ATPases containing two conserved ATPase domains, D1 and D2, preceded by an N-terminal domain that interacts with substrates and adaptor proteins. They assemble to form a heterohexamer in which nucleotide-binding pockets form at the interface between adjacent subunits in the ring [34, 35]. In mammals and yeasts, PEX1 and PEX6 form a heterohexamer that functions as an unfoldase for the retrotranslocation of PEX5 from the peroxisomal membrane [63–65]. Arabidopsis *PEX1*, an ortholog of *S. cerevisiae* Pex1p, was cloned and its cDNA was used to generate transgenic *A. thaliana* harboring *luciferase* under the control of the *PEX1* promoter. The transgenic plants showed that *PEX1* gene expression is upregulated in response to physiological levels of hydrogen peroxide, wounds, and pathogen infection [66]. The Arabidopsis *pex1* mutant *pex1-1* was isolated because of its ability to suppress the *pex6-1* mutant phenotype (Arabidopsis *pex6* mutants are described in the next paragraph) [67]. The *pex1-1* mutant has an amino acid substitution from glutamic acid to lysine at amino acid 748, near the AAA1 domain, and this mutation ameliorates the metabolic and physiological defects of *pex6-1* [67]. Two additional *pex1* mutants, *pex1-2* and *pex1-3*, were isolated from a pool of *ibr* mutants [45]. The mutations in both *pex1*-2 and *pex1-3* are located in the AAA2 domain. The *pex1-2* mutant has reduced PEX1 and PEX6 levels, suggesting that PEX1 enhances PEX6 stability and vice versa [45]. In the *pex1-3* mutant, peroxisomes are larger than those in the wild type, and GFP-PTS1 is detected in the cytosol as well as in peroxisomes; it also displays embryonic lethality [45]. These findings indicate that PEX1 is involved in peroxisome biogenesis and protein transport and that peroxisome-related functions are required for embryogenesis in plants.

The Arabidopsis *pex6-1* mutant was isolated as one of the *ibr m*utants [43]. The *pex6* mutation results in an amino acid substitution from arginine to glutamine at position 766, which is conserved in plants, humans, and yeast. Introduction of human *PEX6* into the Arabidopsis *pex6* mutant functionally complements its phenotype [43]. The *pex6-2* mutant was subsequently isolated as one of the *pfl* (*persistent GFP-ICL fluorescence*) mutants [48]. Isocitrate lyase (ICL) is a peroxisomal protein that is transported via a PTS1-dependent pathway. The *pfl* mutants show a GFP fluorescence pattern that differs from that of the parent plants expressing the *GFP-ICL* fusion gene [48]. Novel alleles of the *pex6* mutant, *pex6-3* and *pex6-4*, were also isolated from the pool of *ibr* mutants [46]. Of the four *pex6* alleles, *pex6-2* has a mutation near the N-terminus of AAA1, and *pex6-1*, *pex6-3*, and *pex6-4* have mutations at or near the AAA2 domain. The mutant phenotypes affecting plant growth and peroxisomal protein transport are similar to those of the wild type in *pex6-2*, whereas *pex6-1*, *pex6-3*, and *pex6-4* show more severe defects, indicating the importance of the AAA2 domain [46]. In PEX1 described in the previous paragraph, the importance of the AAA2 domain is suggested by the fact that of the three Arabidopsis *pex1* mutants, *pex1-2* and *pex1-3,* were isolated as mutants with mutations in the PEX1 AAA2 domain [45, 46].

2.6 Pex15p/PEX26/APEM9 tether the PEX1-PEX6 complex to peroxisome membranes

The ubiquitination of proteins as a signal for protein transport requires energy from ATP hydrolysis on the peroxisomal membrane, which is provided by the PEX1-PEX6 complex. However, PEX1 and PEX6 do not have a membrane association domain. In yeast and mammals, Pex15p and PEX26, which are tail-anchored proteins, are located on the peroxisomal membrane and are involved in recruiting the PEX1-PEX6 complex to the peroxisomal membrane from the cytosol [68, 69]. The homolog of Pex15p/PEX26 in plants was identified after the isolation of the *apem9-1* mutant and based on the sequence of APEM9, the T-DNA insertion mutant was identified as *apem9-2*. Similar to *apem4* and *apem7*, the *apem9* mutant phenotype is characterized by the accumulation of peroxisomal proteins in the cytosol and impaired PTS1- and PTS2-dependent transport (**Figure 3**) [41]. APEM9 encodes a protein of unknown function that is found only in plant genomes; however, hydropathy profile analysis of APEM9 suggests that it is similar to Pex15p and PEX26 [41]. Although APEM9, Pex15p, and PEX26 have low amino acid sequence similarity except in the transmembrane domain, their secondary structures are similar [41]. Analyses suggest that APEM9 tethers the PEX1-PEX6 complex to the peroxisomal membrane, and the *apem9-1* mutation disrupts the peroxisomal localization of APEM9 and the PEX1-PEX6 complex because the mutation is located in the transmembrane domain of APEM9; this suggests that the role of APEM9 is the same as that of Pex15p in yeast and PEX26 in mammals [41]. The Arabidopsis *dayu* mutant is characterized by abnormal pollen maturation and germination [70]. *DAYU* encodes APEM9. DAYU/APEM9 binds to PEX13, a factor of the import complex (**Figure 1**), suggesting that DAYU/APEM9 is involved in the import of both PTS1- and PTS2-containing proteins in addition to mediating receptor export by the ubiquitin system [70]. Moreover, the *pex26-1* mutant, which was isolated as one of the *ibr* mutants [46], shows a more severe phenotype than *apem9-1*. This is likely because the *apem9-1* mutation causes an amino acid substitution, whereas the *pex26-1* mutation triggers a splicing defect that prevents translation of the C-terminal transmembrane domain of PEX26, which in turn prevents the PEX1-PEX6 complex from localizing to the peroxisomal membrane [46].

3. Molecular mechanism underlying the role of ubiquitination in peroxisomal protein transport

Regulatory mechanisms for the ubiquitination of peroxisomal proteins as a signal for transport exist in mammals, yeast, and plants. However, their components and usage differ slightly, and there is no unified model that can be adapted to all organisms. For example, a protein with amino acid sequence homology to PEX4 has not been identified in animals; instead, UbcH5a/b/c act as UBC enzymes [23]. As shown in **Figures 1** and **2**, peroxisomal protein import systems differ between animals, yeast, and plants, suggesting that the mechanism underlying receptor export, the latter step of protein transport triggered by the ubiquitin signal, may also vary from organism to organism. This section outlines the similarities and differences in the regulatory mechanisms according to the components of the ubiquitin system introduced in Section 2, the proteins targeted for ubiquitination, and the protein degradation that occurs when ubiquitin signaling is disrupted by another type of ubiquitination on the peroxisomal membrane.

3.1 Similarities and differences in the proteins involved in the ubiquitin system in animals, yeast, and plants

The presence of three RING-finger ubiquitin ligases, PEX2, PEX10, and PEX12, as E3 enzymes is common to yeast, mammals, and plants. Another common mechanism among the three organisms is the formation of a heterohexamer between PEX1 and PEX6 with AAA ATPase activity; this PEX1-PEX6 complex is recruited to the peroxisomal membrane by a tail-anchored protein: Pex15p in yeast, PEX26 in mammals, and APEM9 in plants [41, 68, 69]. However, the three tail-anchored proteins show low amino acid sequence identity. Indeed, although APEM9 could not be identified by sequence comparison with Pex15p or PEX26, it was identified as a functional homolog of Pex15p and PEX26 in plants because its secondary structure is similar to that of Pex15p and Pex26, especially in the location of the predicted single transmembrane domain and the hydrophilic region consisting of 35–60 residues immediately before the transmembrane domain [41].

There are significant differences between mammals and yeast/plants with regard to the UBC enzyme, namely, the E2 proteins of the ubiquitin system. In yeast and plants, Pex4p and PEX4 serve as E2 enzymes. However, proteins with a similar amino acid sequence to that of Pex4p or PEX4 have not been found in mammals. Instead, three UbcH5-family proteins, UbcH5a, UbcH5b, and UbcH5c, act as UBC enzymes [23]. As discussed in the next section, the peroxisomal protein transport-related target of the UbcH5 family is PEX5; however, the UbcH5 family also targets proteins unrelated to peroxisomes, such as IκBα and BRCA1 [71, 72]. In yeast and plants, Pex4p and PEX4 are UBCs involved only in peroxisomal protein transport, whereas, in animals, UBCs involved in other biological processes also function in peroxisomal protein transport. Proteins corresponding to Pex22p in yeast and PEX22 in plants have not been identified in animals. This is not surprising in animals, where homologs of Pex4p and PEX4 do not exist, and considering that Pex22p and PEX22 are responsible for tethering Pex4p and PEX4 to the peroxisomal membrane.

3.2 Proteins targeted for ubiquitination involved in peroxisomal protein transport

In yeast and mammals, the PTS1 receptor PEX5 is recycled from the peroxisomal membrane to the cytosol by monoubiquitination of the conserved cysteine near the N-terminus [24, 32, 73–75]. However, the final target proteins remain undefined in plants. Although Arabidopsis PEX5 also has a cysteine residue at position 13, there is no direct evidence indicating that PEX5 is recycled via monoubiquitination of the cysteine residue in plants. However, although PEX5 is detected in both the cytosolic and membrane fractions, membrane-associated PEX5 is slightly larger than soluble PEX5 [15]. Furthermore, membrane-associated PEX5 is increased in *apem* and *ibr* mutants, which are characterized by defects in the factors involved in ubiquitination related to peroxisome protein transport [15, 16, 39, 43, 67]. In addition, the mutant phenotype associated with protein transport defects can be partially rescued by overexpression of the *PEX5* gene [43, 48, 67], supporting that PEX5 may be a target for ubiquitination in plants.

In yeast, the PTS2 receptor Pex7p requires a co-receptor such as Pex18p, Pex20p, and Pex21p to function in peroxisomal protein transport. The N-terminal cysteine residue of Pex18p is monoubiquitinated during the recycling process of Pex7p and Pex18p [76]. In mammals, PEX5L, one of the longer splice variants alternatively produced from the *PEX5* gene, functions as a co-receptor for PEX7. The conserved cysteine at the N-terminus of PEX5L is also monoubiquitinated, resulting in the return of both PEX5L and PEX7 to the cytosol from the peroxisomes [77]. As shown in **Figure 1**, in plants, PEX5 and PEX7 are transported as a single complex after recognizing PTS1 and PTS2, respectively, suggesting that if PEX5 is ubiquitinated and recycled to the cytosol, PEX7 is also returned, similar to PEX5L and PEX7 in mammals.

3.3 A different ubiquitination process on the peroxisomal membrane

Abnormal peroxisomal protein transport is associated with a type of ubiquitination on the peroxisomal membrane that differs from the monoubiquitination of PEX5 and Pex18p. In yeast, inhibition of monoubiquitination-dependent protein transport induces the polyubiquitination of Pex5p [22, 78, 79]. In this case, ubiquitin is conjugated to the 2 N-terminal lysine residues, but not to the cysteine residue, by another UBC enzyme, Ubc4p (and the partially redundant UBCs, Ubc1p, and Ubc5p) rather than Pex4p [22, 78, 79]. Polyubiquitinated Pex5p is then degraded by the 26S proteasome. With regard to PTS2-dependent protein transport, Pex18p and Pex20p, co-receptors for Pex7p, are lysine polyubiquitinated and degraded [76, 80]. In mammals, dysfunctional PEX7 is degraded via a ubiquitin-dependent pathway for PEX7 quality control [81].

In plants, the ubiquitination-dependent degradation of PEX5 and PEX7 has not been identified to date. However, Arabidopsis PEX4 catalyzes the formation of lysine48-linked ubiquitin chains [16], suggesting that PEX4 is involved in the degradation as well as recycling of PEX5. Dominant suppressors of KAR 2a, DSK2a and DSK2b, are ubiquitin-binding receptor proteins that specifically bind to the RING domain of PEX2 and PEX12, suggesting that the E3 ligases associated with DSK2s are involved in ubiquitination on the peroxisomal membrane [33].

Suppressors of plastid protein import locus 1 (SP1) and SP1-like 1 (SPL1) are RING-type ubiquitin ligases that are located on the peroxisomal membrane [82, 83]. SP1 physically interacts with PEX13 and PEX14, components of the import machinery, and with PEX2, thereby contributing to the destabilization of the protein import machinery [82]. However, SP1 was originally discovered as a component of the chloroplast protein import machinery and is involved in the degradation of the chloroplast protein transport complex by the ubiquitin-proteasome system [84]. SP1 localizes to peroxisomes and mitochondria as well as chloroplasts [84, 85]. SPL1 is the closest homolog of SP1, and similar to SP1, localizes to chloroplasts, mitochondria, and peroxisomes; however, SP1 shows a stronger peroxisomal localization than SPL1 [83]. SPL1 antagonizes SP1 function during peroxisomal protein import, suggesting that the balance of SP1 and SPL1 activities regulates the protein import machinery [83]. Thus, ubiquitination is required for the import of peroxisomal proteins as well as the export of receptors in the protein transport process. A protein with amino acid sequence similarity to SP1 and SPL1 is the human mitochondrial anchored protein ligase (MAPL) protein [86]. MAPL has not been identified in yeast. In mitochondria, it functions in mitochondrial fission by regulating dynamin-associated protein 1. Whether it functions in peroxisomes such as SP1 and SPL1 as potential regulators of docking proteins remains unknown [86]. Peroxisomes interact with multiple organelles, such as chloroplasts and mitochondria, to form metabolic pathways. Therefore, E3 ligases localized to multiple organelles may play an important role in coordinately regulating processes related to organelle biogenesis, such as protein transport, by recognizing interactions between organelles.

4. Conclusions

Proteins involved in ubiquitination on the peroxisomal membrane have been identified in plants, and their molecular mechanisms have been elucidated. As described above, the regulatory mechanisms in plants are similar to those in animals and yeast, but with species-specific fine-tuning that may have been acquired during evolution. Disruption of this ubiquitin system can lead to abnormalities in plant processes such as growth and embryogenesis, similar to the inability of yeast to use carbon sources for growth and the occurrence of genetic diseases in humans.

Several issues related to the role of ubiquitin signaling in peroxisomal protein transport in plants remain to be addressed. Proteins targeted for ubiquitination in this system remain to be identified in plants. Whether PEX5 is ubiquitinated as in plants is a subject for further study. In yeast and animals, ubiquitinated PEX5 is released from the peroxisomal membrane to the cytosol, where it is deubiquitinated before engaging in the following transport cycle. The enzyme responsible for PEX5 deubiquitination is ubiquitin-specific protease 15 (Ubp15p) in yeast and ubiquitin-specific protease 9X (USP9X) in mammals [87, 88]; however, the deubiquitinase in plants has not been identified. Moreover, the mechanism underlying polyubiquitination-induced protein degradation on the peroxisomal membrane, when protein transport fails, is not as well understood in plants as in yeast and animals, including whether a similar mechanism exists in the first place. Next-generation sequencing and mass spectrometry will contribute to the identification of ubiquitin-related factors and ubiquitinated proteins in various nonmodel plants. Analysis using a variety of plants, including the liverwort *Marchantia polymorpha*, mentioned in the next section, will facilitate the elucidation of the molecular mechanisms of the ubiquitin system in plants and help clarify the issues discussed above.

5. Future perspectives

The current knowledge of peroxisomal protein transport in land plants presented in this chapter is based primarily on studies using Arabidopsis. However, these findings may not apply to all plants. In yeast, although the ubiquitin system involved in peroxisomal protein transport has been preserved through the course of evolution, the factors involved and their functions have changed; for example, proteins from *S. cerevisiae* do not complement their corresponding mutants in *Pichia pastoris*. Similarly, ubiquitin factors may have diverged during plant evolution. In this regard, the liverwort *M. polymorpha* is a good experimental plant for the study of peroxisomes. *M.polymorpha* is an early diverging land plant and thus retains features of ancestral land plants. The genome of *M. polymorpha* has been determined and compiled into genome databases (https://marchantia.info) [89]. Methods for genetic transformation, as well as various resources such as vectors suitable for *M. polymorpha* and genome editing by CRISPR/Cas9 have been established, which has enabled the collection of information on different mutants [90]. One of the advantages of using *M. polymorpha* is that it has a low gene duplication rate. The details of this process are beyond the scope of this chapter, and several reviews are available [12, 90, 91]. We have successfully visualized the peroxisome of *M. polymorpha* using a fusion gene of PTS1 or PTS2 and a fluorescent protein [12, 92]. The results show that both PTS1- and PTS2-dependent protein transport have been present from the beginning of the evolution of land plants. Bioinformatics analysis detected the presence of genes orthologous to Arabidopsis *PEX1*, *PEX2*, *PEX4*, *PEX6*, *PEX10*, *PEX12*, *PEX22*, and *APEM9* in the genome of *M. polymorpha* [12], indicating that the ubiquitin system is involved in peroxisomal protein transport in *M. polymorpha* [the *PEX* genes we have identified, including the ubiquitin system-related *PEX* genes, are registered in MarpolBase (https://marchantia.info)]. However, the molecular mechanisms underlying the roles of the gene products need to be analyzed. Genome editing of ubiquitin-related genes identified by bioinformatics analysis of *M. polymorpha* transgenic plants with visualized peroxisomes will help clarify the ubiquitin signaling pathways regulating protein transport to peroxisomes in land plants.

Acknowledgements

We thank the staff at the Model Organisms Facility, Trans-Omics Facility, Optics, and Imaging Facility at the NIBB Trans-Scale Biology Center for technical support. We are also grateful to Ms. Chihiro Nakamori, Masami Araki, and Azusa Matsuda for supporting the experiments and taking care of the plants as technical staff in our laboratory.

This work was supported in part by JSPS KAKENHI (Grant Numbers 20059035, 22112523, 17 K07457, and 20 K06711).

Author contribution

Shoji Mano, Conceptualization; Investigation, Writing manuscript, Project administration, Funding acquisition

Kazumi Hikino, Investigation

Masatake Kanai, Investigation, Writing manuscript

Conflicts of interest

The authors have no conflicts of interest to declare.

Abbreviations

apem	aberrant peroxisome morphology
AWP1	associated with PRK1
BRCA1	breast cancer susceptibility gene 1
CRISPR/Cas9	clustered regularly interspaced short palindromic repeats/CRISPR-associated proteins
dayu	*dau*, after the Chinese legendary hero
det1	*de-etiolated 1*
DSK2a	dominant suppressor of KAR 2a
EMS	ethyl methanesulfonate
GFP	green fluorescent protein
HECT	homologous to the E6-AP carboxyl terminus
IAA	indole-3-acetic acid
IBA	indole-3-butyric acid
ibr	*iba-response*
ICL	isocitrate lyase
IκBα	NF-κB inhibitor α
MAPL	mitochondrial anchored protein ligase
pfl	*persistent GFP-ICL fluorescence*
RBR	RING-in-between-RING
PEX	peroxin
PTS	peroxisome targeting signal
RING	really interesting new gene
SP1	plastid protein import locus 1

ted	reversal of the *det* phenotype
UBC	ubiquitin conjugating
Ubp15p	ubiquitin-specific protease 15
USP9X	ubiquitin-specific protease 9X
AAA ATPase	ATPases associated with diverse cellular activities

Author details

Shoji Mano[1,2]*, Kazumi Hikino[1] and Masatake Kanai[1]

1 Laboratory of Organelle Regulation, National Institute for Basic Biology, Okazaki, Japan

2 Basic Biology Program, Graduate Institute for Advanced Studies, The Graduate University for Advanced Studies, SOKENDAI, Okazaki, Japan

*Address all correspondence to: mano@nibb.ac.jp

References

[1] Rhodin J. Anatomy of kidney tubules. International Review of Cytology. 1958;7:485-534. DOI: 10.1016/S0074-7696(08)62694-7

[2] Sakai Y, Oku M, Klei IJvd, Kiel JAKW. Pexophagy: Autophagic degradation of peroxisomes. Biochimica et Biophysica Acta. 2006;**1763**:1767-1775. DOI: 10.1016/j.bbamcr.2006.08.023

[3] Shibata M, Oikawa K, Yoshimo K, Kondo M, Mano S, et al. Highly oxidized peroxisomes are selectively degraded via autophagy in *Arabidopsis*. The Plant Cell. 2013;**25**:4967-7983. DOI: 10.1105/tpc.113.116947

[4] Yoshimoto K, Shibata M, Kondo M, Oikawa K, Sato M, et al. Organ-specific quality control of plant peroxisomes is mediated by autophagy. Journal of Cell Science. 2014;**127**:1161-1168. DOI: 10.1242/jcs.139709

[5] Young PG, Bartel B. Pexophagy and peroxisomal protein turnover in plants. Biochimica et Biophysica Acta - Molecular Cell Research. 2016;**1863**:999-1005. DOI: 10.1016/j.bbamcr.2015.09.005

[6] Eberhart T, Kovac WJ. Pexophagy in yeast and mammals: An update on mysteries. Histochemistry and Cell Biology. 2018;**150**:473-488. DOI: 10.1007/s00418-018-1724-3

[7] Oikawa K, Goto-Yamada S, Hayashi Y, Takahashi D, Kimori Y, et al. Pexophagy suppresses ROS-induced damage in leaf cells under high-intensity light. Nature Communications. 2022;**13**:7493. DOI: 10.1038/s41467-022-35138-z

[8] Nito K, Kamigaki A, Kondo M, Hayashi M, Nishimura M. Functional classification of *Arabidopsis* peroxisome biogenesis factors proposed from analyses of knockdown mutants. Plant and Cell Physiology. 2007;**48**:763-774. DOI: 10.1093/pcp/pcm053

[9] Hu J, Baker A, Bartel B, Linka N, Mullen RT, et al. Plant peroxisomes: Biogenesis and function. The Plant Cell. 2012;**24**:2279-2303. DOI: 10.1105/tpc.112.096586

[10] Fujiki Y, Abe Y, Imoto Y, Tanaka AJ, Okumoto K, et al. Recent insights into peroxisome biogenesis and associated diseases. Journal of Cell Science. 2020;**133**:236943. DOI: 10.1242/jcs.236943

[11] Jansen RLM, Santana-Molina C, van den Noort M, Devos DP, van der Klei IJ. Comparative genomics of peroxisome biogenesis proteins: Making sense of the PEX proteins. Frontiers in Cell and Developmental Biology. 2021;**9**:654163. DOI: 10.3389/fcell.2021.654163

[12] Goto-Yamada S, Oikawa K, Yamato KT, Kanai M, Hikino K, et al. Image-based analysis revealing the molecular mechanism of peroxisome dynamics in plants. Frontiers in Cell and Developmental Biology. 2022;**10**:883491. DOI: 10.3389/fcell.2022.883491

[13] Francisco T, Rodrigues TA, Pinto MP, Carvalho AF, Azevedoa JE, et al. Ubiquitin in the peroxisomal protein import pathway. Biochimie. 2014;**98**:29-35. DOI: 10.1016/j.biochi.2013.08.003

[14] Platta HW, Brinkmeier R, Reidick C, Galiani S, Clausen MP, et al. Regulation of peroxisomal matrix protein import by ubiquitination. Biochimica et Biophysica Acta. 2016;**1863**:838-849. DOI: 10.1016/j.bbamcr.2015.09.010

[15] Mano S, Hayashi Y, Hikino K, Otomo M, Kanai M, et al.

Ubiquitin-conjugating activity by PEX4 is required for efficient protein transport to peroxisomes in *Arabidopsis thaliana*. Journal of Biological Chemistry. 2022;**298**:102038. DOI: 10.1016/j.jbc.2022.102038

[16] Traver MS, Bradford SE, Olmos JL, Wright ZJ, Miller MD, et al. The structure of the *Arabidopsis* PEX4-PEX22 peroxin complex-insights into ubiquitination at the peroxisomal membrane. Frontiers in Cell and Developmental Biology. 2022;**10**:838923. DOI: 10.3389/fcell.2022.838923

[17] Kato A, Hayashi M, Mori H, Nishimura M. Molecular characterization of a glyoxysomal citrate synthase that is synthesized as a precursor of higher molecular mass in pumpkin. Plant Molecular Biology. 1995;**27**:377-390. DOI: 10.1007/BF00020191

[18] Kato A, Hayashi M, Takeuchi Y, Nishimura M. cDNA cloning and expression of a gene for 3-ketoacyl-CoA thiolase in pumpkin cotyledons. Plant Molecular Biology. 1996;**31**:843-852. DOI: 10.1007/BF00019471

[19] Kato A, Takeda-Yoshikawa Y, Hayashi M, Kondo M, Hara-Nishimura I, et al. Glyoxysomal malate dehydrogenase in pumpkin: Cloning of a cDNA and functional analysis of its presequence. Plant and Cell Physiology. 1998;**39**:186-195. DOI: 10.1093/oxfordjournals.pcp.a029356

[20] Purdue PE, Yang X, Lazarow PB. Pex18p and Pex21p, a novel pair of related peroxins essential for peroxisomal targeting by the PTS2 pathway. Journal of Cell Biology. 1998;**143**:1859-1869. DOI: 10.1083/jcb.143.7.1859

[21] Léon S, Subramani S. A conserved cysteine residue of *Pichia pastoris* Pex20p is essential for its recycling from the peroxisome to the cytosol. Journal of Biological Chemistry. 2007;**282**:7424-7430. DOI: 10.1074/jbc.M611627200

[22] Platta HW, Girzalsky W, Erdmann R. Ubiquitination of the peroxisomal import receptor Pex5p. Biochemical Journal. 2004;**384**:37-45. DOI: 10.1042/BJ20040572

[23] Grou CP, Carvalho AF, Pinto MP, Huybrechts SJ, Sá-Miranda C, et al. Properties of the ubiquitin-Pex5p thiol ester conjugate. Journal of Biological Chemistry. 2009;**284**:10504-10513. DOI: 10.1074/jbc.M808978200

[24] Okumoto K, Misono S, Miyata N, Matsumoto Y, Mukai S, et al. Cysteine ubiquitination of PTS1 receptor Pex5p regulates Pex5p recycling. Traffic. 2011;**12**:1067-1083. DOI: 10.1111/j.1600-0854.2011.01217.x

[25] Miyata N, Okumoto K, Mukai S, Noguchi M, Fujiki Y. AWP1/ZFAND6 functions in Pex5 export by interacting with Cys-monoubiquitinated Pex5 and Pex6 AAA ATPase. Traffic. 2012;**13**:168-183. DOI: 10.1111/j.1600-0854.2011.01298.x

[26] Hayashi M, Yagi M, Nito K, Kamada T, Nishimura M. Differential contribution of two peroxisomal protein receptors to the maintenance of peroxisomal functions in *Arabidopsis*. Journal of Biological Chemistry. 2005;**280**:14829-14835. DOI: 10.1074/jbc.M411005200

[27] Wiebel FF, Kunau W-H. The Pas2 protein essential for peroxisome biogenesis is related to ubiquitin-conjugating enzymes. Nature. 1992;**359**:73-76. DOI: 10.1038/359073a0

[28] Crane DI, Kalish JE, Gould SJ. The *Pichia pastoris PAS4* gene encodes a ubiquitin-conjugating enzyme required for peroxisome assembly.

Journal of Biological Chemistry. 1994;**269**:21835-21844. DOI: 10.1016/S0021-9258(17)31879-3

[29] Klei IJVD, Hilbrands RE, Kiel JAKW, Rasmussen SW, Cregg JM, et al. The ubiquitin-conjugating enzyme Pex4p of *Hansenula polymorpha* is required for efficient functioning of the PTS1 import pathway. The EMBO Journal. 1998;**17**:3608-3618. DOI: 10.1093/emboj/17.13.3608

[30] Zolman BK, Monroe-Augustus M, Silva ID, Bartel B. Identification and functional characterization of *Arabidopsis* PEROXIN4 and the interacting protein PEROXIN22. The Plant Cell. 2005;**17**:3422-3435. DOI: 10.1105/tpc.105.035691

[31] Williams C, Mvd B, Geers E, Distela B. Pex10p functions as an E3 ligase for the Ubc4p-dependent ubiquitination of Pex5p. Biochemical and Biophysical Research Communications. 2008;**374**:620-624. DOI: 10.1016/j.bbrc.2008.07.054

[32] Platta HW, Magraoui FE, Bäumer BE, Schlee D, Girzalsky W, et al. Pex2 and Pex12 function as protein-ubiquitin ligases in peroxisomal protein import. Molecular and Cellular Biology. 2009;**29**:5505-5516. DOI: 10.1128/MCB.00388-09

[33] Kaur N, Zhao Q, Xie Q, Hu J. *Arabidopsis* RING peroxins are E3 ubiquitin ligases that interact with two homologous ubiquitin receptor proteins. Journal of Integrative Plant Biology. 2013;**55**:108-120. DOI: 10.1111/jipb.12014

[34] Saffian D, Grimm I, Girzalsky W, Erdmann R. ATP-dependent assembly of the heteromeric Pex1p–Pex6p-complex of the peroxisomal matrix protein import machinery. Journal of Structural Biology. 2012;**179**:126-132. DOI: 10.1016/j.jsb.2012.06.002

[35] Gardner BM, Chowdhury A, Lander GC, Martin A. The Pex1/Pex6 complex is a heterohexameric AAA + motor with alternating and highly coordinated subunits. Journal of Molecular Biology. 2015;**427**:1375-1388. DOI: 10.1016/j.jmb.2015.01.019

[36] Koller A, Snyder WB, Faber KN, Wenzel TJ, Rangell L, et al. Pex22p of *Pichia pastoris*, essential for peroxisomal matrix protein import, anchors the ubiquitin-conjugating enzyme, Pex4p, on the peroxisomal membrane. The Journal of Cell Biology. 1999;**146**:99-[illegible] DOI: 10.1083/jcb.146.1.99

[37] Collins CS, Kalish JE, Morrell JC, McCaffery JM, Gould SJ. The peroxisome biogenesis factors Pex4p, Pex22p, Pex1p, Pex6p act in the terminal steps of peroxisomal matrix protein import. Molecular and Cellular Biology. 2000;**20**:7516-7526. DOI: 10.1128/MCB.20.20.7516-7526.2000

[38] Mano S, Nakamori C, Kondo M, Hayashi M, Nishimura M. An *Arabidopsis* dynamin-related protein, DRP3A, controls both peroxisomal and mitochondrial division. The Plant Journal. 2004;**38**:487-498. DOI: 10.1111/j.1365-313X.2004.02063.x

[39] Mano S, Nakamori C, Nito K, Kondo M, Nishimura M. The Arabidopsis *pex12* and *pex13* mutants are defective in both PTS1- and PTS2-dependent protein transport to peroxisomes. The Plant Journal. 2006;**47**:604-618. DOI: 10.1111/j.1365-313X.2006.02809.x

[40] Mano S, Nakamori C, Fukao Y, Araki M, Matsuda A, et al. A defect of peroxisomal membrane protein 38 causes enlargement of peroxisomes. Plant and Cell Physiology. 2011;**52**:2157-2172. DOI: 10.1093/pcp/pcr147

[41] Goto S, Mano S, Nakamori C, Nishimura M. *Arabidopsis* aberrant

peroxisome morphology9 is a peroxin that recruits the PEX1-PEX6 complex to peroxisomes. The Plant Cell. 2011;**23**:1573-1587. DOI: 10.1105/tpc.110.080770

[42] Goto-Yamada S, Mano S, Nakamori C, Kondo M, Yamawaki R, et al. Chaperone and protease functions of LON protease 2 modulate the peroxisomal transition and degradation with autophagy. Plant and Cell Physiology. 2014;**55**:482-496. DOI: 10.1093/pcp/pcu017

[43] Zolman BK, Bartel B. An *Arabidopsis* indole-3-butyric acid-response mutant defective in PEROXIN6, an apparent ATPase implicated in peroxisomal function. Proceedings of the National Academy of Sciences of The United States of America. 2004;**101**:1786-1791. DOI: 10.1073/pnas.0304368101

[44] Kao YT, Fleming WA, Ventura MJ, Bartel B. Genetic interactions between PEROXIN12 and other peroxisome-associated ubiquitination components. Plant Physiology. 2016;**172**:1643-1656. DOI: 10.1104/pp.16.01211

[45] Rinaldi MA, Fleming WA, Gonzalez KL, Park J, Ventura MJ, et al. The PEX1 ATPase stabilizes PEX6 and plays essential roles in peroxisome bology. Plant Physiology. 2017;**174**:2231-2247. DOI: 10.1104/pp.17.00548

[46] Gonzalez KL, Fleming WA, Kao YT, Wright ZJ, Venkova SV, et al. Disparate peroxisome-related defects in Arabidopsis *pex6* and *pex26* mutants link peroxisomal retrotranslocation and oil body utilization. The Plant Journal. 2017;**92**:110-128. DOI: 10.1111/tpj.13641

[47] Hu J, Aguirre M, Peto C, Alonso J, Ecker J, et al. A role for peroxisomes in photomorphoenesis and development of *Arabidopsis*. Science. 2002;**297**:405-409. DOI: 10.1126/science.1073633

[48] Burkhart SE, Lingard MJ, Bartel B. Genetic dissection of peroxisome-associated matrix protein degradation in *Arabidopsis thaliana*. Genetics. 2013;**193**:125-141. DOI: 10.1534/genetics.112.146100

[49] Kao YT, Bartel B. Elevated growth temperature decreases levels of the PEX5 peroxisome-targeting signal receptor and ameliorates defects of Arabidopsis mutants with an impaired PEX4 ubiquitin-conjugating enzyme. BMC Plant Biology. 2015;**15**:224. DOI: 10.1186/s12870-015-0605-3

[50] Magraoui FE, Schrötter A, Brinkmeier R, Kunst L, Mastalski T, et al. The cytosolic domain of Pex22p stimulates the Pex4p-dependent ubiquitination of the PTS1-receptor. PLoS One. 2014;**9**:e105894. DOI: 10.1371/journal.pone.0105894

[51] Williams C, Mvd B, Panjikar S, Stanley WA, Distel B, et al. Insights into ubiquitin-conjugating enzyme/co-activator interactions from the structure of the Pex4p:Pex22p complex. The EMBO Journal. 2012;**31**:391-402. DOI: 10.1038/emboj.2011.411

[52] Williams C, van den Berg M, Stanley WA, Wilmanns M, Distel B. A disulphide bond in the E2 enzyme Pex4p modulates ubiquitin-conjugating activity. Scientific Reports. 2013;**3**:2212. DOI: 10.1038/srep02212

[53] Hazra PP, Suriapranata I, Snyder WB, Subramani S. Peroxisome remnants in pex3delta cells and the requirement of Pex3p for interactions between the peroxisomal docking and translocation subcomplexes. Traffic. 2002;**3**:560-574. DOI: 10.1034/j.1600-0854.2002.30806.x

[54] Agne B, Meindl NM, Niederhoff K, Einwächter H, Rehling P, et al. Pex8p: An intraperoxisomal organizer of the peroxisomal import machinery. Molecular Cell. 2003;**11**:635-646. DOI: 10.1016/s1097-2765(03)00062-5

[55] Burkhart SE, Kao Y-T, Bartel B. Peroxisomal ubiquitin-protein ligases Peroxin2 and Peroxin10 have distinct but synergistic roles in matrix protein import and Peroxin5 retrotranslocation in Arabidopsis. Plant Physiology. 2014;**166**:1329-1344. DOI: 10.1104/pp.114.247148

[56] Schwechheimer C, Deng XW. The COP/DET/FUS proteins-regulators of eukaryotic growth and development. Seminars in Cell and Developmental Biology. 2000;**11**:495-503. DOI: 10.1006/scdb.2000.0203

[57] Desai M, Kaur N, Hu J. Ectopic expression of the RING domain of the Arabidopsis Peroxin2 protein partially suppresses the phenotype of the photomorphogenic mutant *de-etiolated1*. PLoS One. 2014;**9**:e108473. DOI: 10.1371/journal.pone.0108473

[58] Sparkes IA, Brandizzi F, Slocombe SP, El-Shami M, Hawes C, et al. An Arabidopsis *pex10* null mutant is embryo lethal, implicating peroxisomes in an essential role during plant embryogenesis. Plant Physiology. 2003;**133**:1809-1819. DOI: 10.1104/pp.103.031252

[59] Schumann U, Wanner G, Veenhuis M, Schmid M, Gietl C. AthPEX10, a nuclear gene essential for peroxisome and storage organelle formation during *Arabidopsis* embryogenesis. Proceedings of the National Academy of Sciences of The United States of America. 2003;**100**:9626-9631. DOI: 10.1073/pnas.1633697100

[60] Kamigaki A, Kondo M, Mano S, Hayashi M, Nishimura M. Suppression of peroxisome biogenesis factor 10 reduces cuticular wax accumulation by disrupting the ER network in *Arabidopsis thaliana*. Plant and Cell Physiology. 2009;**50**:2034-2046. DOI: 10.1093/pcp/pcp152

[61] Fan J, Quan S, Orth T, Awai C, Chory J, et al. The Arabidopsis *PEX12* gene is required for peroxisome biogenesis and is essential for development. Plant Physiology. 2005;**139**:231-239. DOI: 10.1104/pp.105.066811

[62] Singh T, Hayashi M, Mano S, Arai Y, Goto S, et al. Molecular components required for the targeting of PEX7 to peroxisomes in *Arabidopsis thaliana*. The Plant Journal. 2009;**60**:488-498. DOI: 10.1111/j.1365-313X.2009.03970.x

[63] Ciniawsky S, Grimm I, Saffian D, Girzalsky W, Erdmann R, et al. Molecular snapshots of the Pex1/6 AAA+ complex in action. Nature Communications. 2015;**6**:7331. DOI: 10.1038/ncomms8331

[64] Gardner BM, Castanzo DT, Chowdhury S, Stjepanovic G, Stefely MS, et al. The peroxisomal AAA-ATPase Pex1/Pex6 unfolds substrates by processive threading. Nature Communications. 2018;**9**:135. DOI: 10.1038/s41467-017-02474-4

[65] Pedrosa AG, Francisco T, Bicho D, Dias AF, Barros-Barbosa A, et al. Peroxisomal monoubiquitinated PEX5 interacts with the AAA ATPases PEX1 and PEX6 and is unfolded during its dislocation into the cytosol. Journal of Biological Chemistry. 2018;**293**:11553-11563. DOI: 10.1074/jbc.RA118.003669

[92] Mano S, Nishihama R, Ishida S, Hikino K, Kondo M, et al. Novel gateway binary vectors for rapid tripartite DNA assembly and promoter analysis with various reporters and tags in the liverwort *Marchantia polymorpha*. PLoS One. 2018;**13**:e0204964. DOI: 10.1371/journal.pone.0204964

Chapter 3

Prenylated Proteins: Structural Diversity and Functions

Aravind Kamath and Kantharaju Kamanna

Abstract

The cell membranes consist of lipid bilayers that are semipermeable. The semipermeable nature enables the cell membranes to regulate the transport of materials entering and exiting the cell. Apart from providing protection and a fixed environment to the cell, the cell membrane has several functions. The covalently linked proteins to lipids on the surface of the cell membranes are the Lipid-anchored proteins. The function of the protein to which the lipid is attached depends on the type of the lipid. Prenylated proteins, fatty acylated proteins, and glycosylphosphatidylinositol-linked proteins (GPI) are the three main types of lipid-anchored proteins on the cell membrane. In particular, the prenylated proteins are very important for cell growth, differentiation, and morphology. The dynamic interaction of prenylated proteins with the cell membrane is important for their signaling functions and is often deregulated in disease processes, such as cancer. An understanding of the prenylated proteins and their mechanisms is important for drug development efforts to combat cancer.

Keywords: Prenylation, lipid-anchored proteins, cell signaling, cancer, drug development

1. Introduction

The diversity in living organisms is primarily characterized by the proteins found in them. Proteins are large, complex, and diverse in their nature. Their abundance is approximately 50% of the dry tissue weight of the organism. Proteins play several critical functional roles in the body, like providing the structural basis, functional, and regulatory activities in the body's cells, tissues, and organs. Proteins form a very important part of the cell membranes, acting as receptors that trigger specific responses in cells. The semipermeable lipid bilayer of the cell membranes enables the regulation of the materials entering and exiting the cell. On either side of the cell membranes, we can find the lipid-anchored proteins. The lipid-anchored proteins on the surface of the cell membranes have diversity in their function depending not only on the type of the lipid covalently attached to the protein but also on the site where the lipid binds to the protein. The three main types of lipid anchored proteins on the cell membrane are prenylated proteins, fatty acylated proteins, and glycosylphosphatidylinositol-linked proteins (GPI). These proteins play a very important role in cell interactions due to the diversity of the lipid groups attached to them. The lipid groups

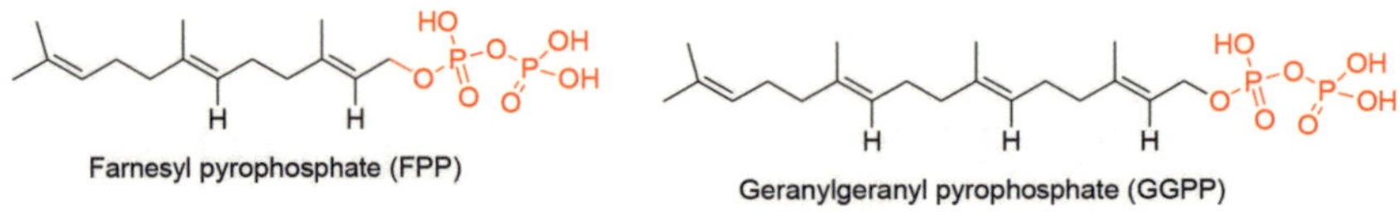

Figure 1.
Prenylation substrates.

alter the molecular hydrophobicity and allow the interaction of cell membranes and protein domains with the lipid-anchored proteins [1].

Proteins are formed as a result of the translation of the nucleic acid sequence to polypeptide sequence. But this simple translation is insufficient to cater to the ever-expanding need for the protein functions necessitated by the complex environmental stimuli [2]. The organisms have thus evolved a vast repertoire of co-translational and post-translational modifications to the protein architecture. Prenylation and/or lipidation are the major post-translational modifications that control the localization and activity of several proteins having crucial regulatory functions. Critical changes in the structure and physicochemical properties of proteins are brought about by the attachment of the lipid groups to the protein. This modification affects the cellular localization, stability, and biological activity of the modified proteins. The lipid modifications in proteins were first discovered by Folch in 1951 [3]. A mating factor in fungus *Rhodospiridium toruloides* was found to be S-farnesylated peptide, a class of prenylated protein in 1978 [4]. Nuclear lamin proteins namely, B lamins and prelamin A were the first prenylated proteins that were identified in mammalian systems [5, 6]. Since then several proteins and peptides known to be modified with prenylation are reported and their number is increasing. Prenylation has been studied extensively owing to its role in the function of the proteins in several cellular activities. Prenylation of proteins include the farnesylation and geranylgeranylation, with the farnesyl pyrophosphate (FPP) or geranylgeranyl pyrophosphate (GGPP) as substrates respectively (**Figure 1**). These hydrophobic prenyl groups anchor the proteins to the cellular membranes and can trigger several cells signaling pathways.

2. Prenylation: its mechanism

The five-carbon (C5) isopentenyl pyrophosphate (IPP) and its isomer dimethylallyl pyrophosphate (DMAPP) (**Figure 2**) are termed as isoprene units are the building blocks of isoprenoids also called as terpenoids. These lipophilic molecules are ubiquitous and exist in all living organisms. They represent the largest and most diverse group of natural compounds [7]. Thousands of isoprenoids are synthesized by

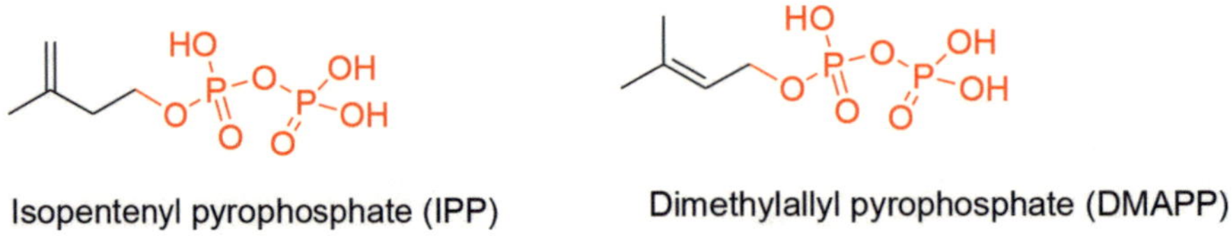

Figure 2.
Building blocks of isoprenoids.

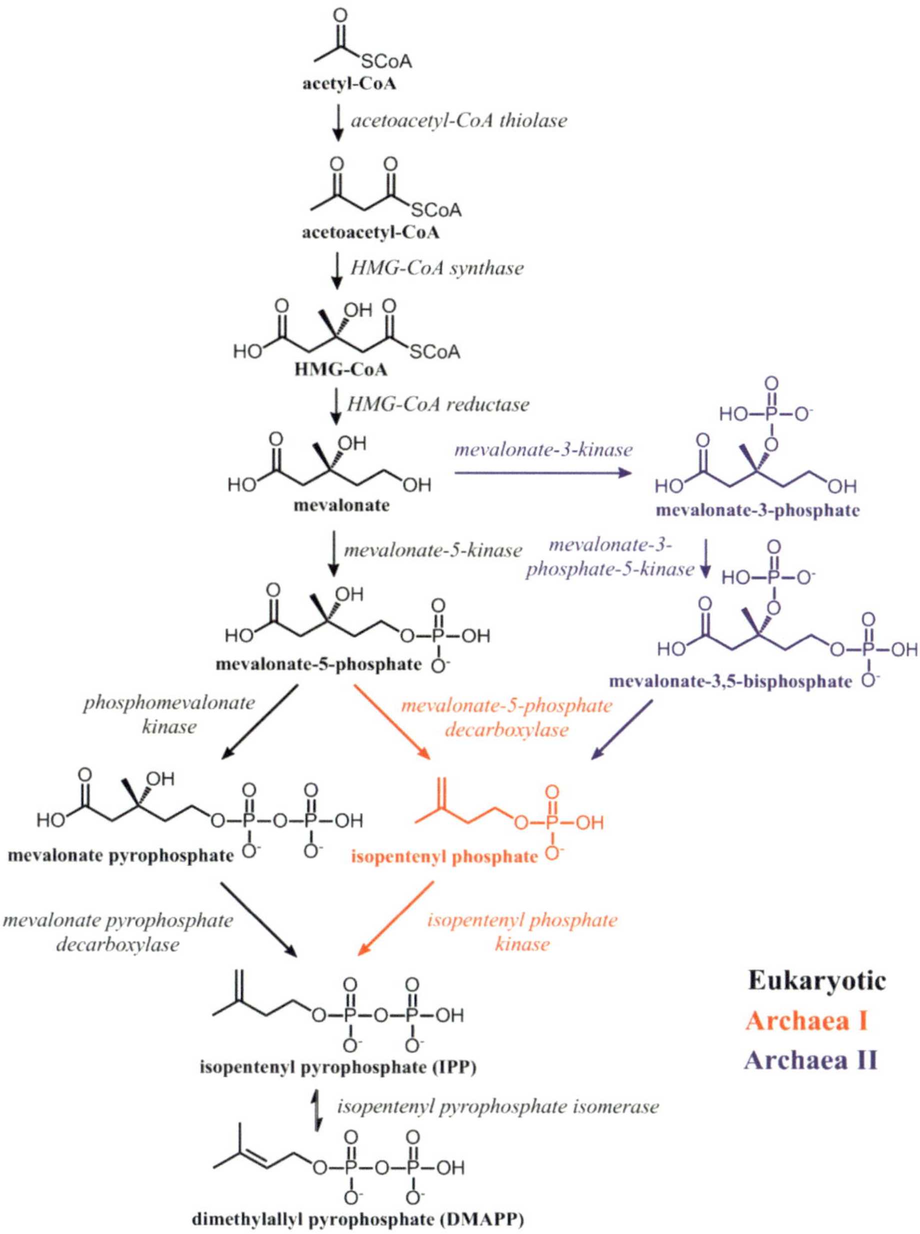

Figure 3.
Mevalonate pathway. This file is licensed under the Creative Commons Attribution-Share Alike 4.0 International.

"head to tail" or "head to head" addition of either IPP or DMAPP. All living organisms can synthesize isoprenoids. But the mevalonate (MVA) pathway (**Figure 3**) is exclusively used by animals, fungi, and archaebacteria for the synthesis of isoprenoids [8]. A series of reactions beginning with the conversion of acetyl-CoA to MVA and further to IPP and DMAPP constitutes the MVA pathway. With the synthesis of IPP and DMAPP, further head to tail condensation of two IPP molecules with DMAPP gives FPP (C15) catalyzed by Farnesyl pyrophosphate synthase (FPS; EC 2.5.1.10).

Subsequent addition of another molecule of IPP to FPP catalyzed by GeranylGeranyl diphosphate synthase (GGPPS; EC 2.5.1.29) gives GGPP(C20). The enzymes involved in the conversion of acetyl-CoA to IPP and DMAPP have been identified in peroxisomes [9] as well as in the cytosol [10].

The biosynthesis of the isoprenoids was once thought to occur exclusively by the MVA pathway. But the identification of a non-MVA pathway in plants and micro-organisms was suggested by the experimental data and this pathway was named as methylerythritol 4-phosphate pathway (MEP) after the precursor (**Figure 4**) [11, 12]. While most eubacteria and *Plasmodium falciparaum*, the malarial parasite uses the MEP pathway, archaebacteria, fungi, and animals use the MVA pathway to synthesize the isoprenoids. It is the plants that use both the pathways albeit in different compartments [13, 14]. Since the majority of the pathogenic bacteria employ the MEP pathway and the same is missing in the mammalian system, the enzymes of the MEP pathway are potent targets for the new anti-infective drugs [15]. In recent years following the discovery that FPP and GGPP could modify the structure and function of unique groups of proteins, which are implicated in cancer and other disorders, through the process of protein prenylation, the interest in isoprenoids and related pathways has increased substantially [16].

The attachment of a lipophilic isoprenoid group covalently to a molecule is referred to as prenylation. FPP and GGPP, generally referred to as prenyl groups are the most relevant isoprenoids relevant to protein prenylation. Protein prenylation is an irreversible covalent post-translational modification comprising of the addition of FPP or GGPP covalently to the cysteine residue at the C- terminus of the protein. Protein prenylation through a thioether linkage (C-S-C) is a relatively stable bond. The cysteine residue located at the C-terminus consists of "CAAX" consensus sequence, with C being cysteine, A being any aliphatic amino acid except alanine, and X being serine, alanine, methionine, or glutamine in case of farnesylation and X being leucine or isoleucine, in case of geranylgeranylation. Either farnesylation or geranylgeranylation

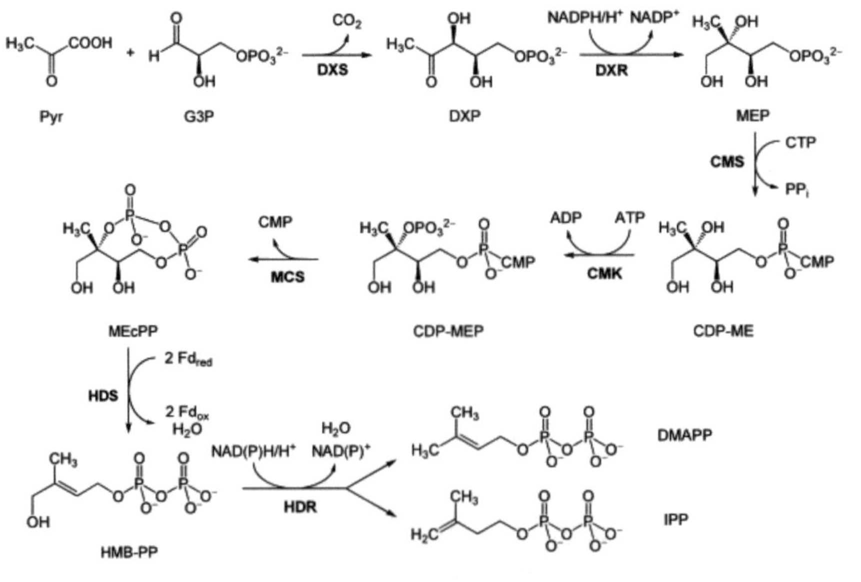

Figure 4.
MEP pathway reactions in the biosynthesis of isoprenoids. Redrawn verbatim from the scheme of Qidwai and coworkers "Exploring Drug Targets in Isoprenoid Biosynthetic Pathway for Plasmodium falciparum." Biochemistry Research International. 2014: 657189.

C terminal consensus sequence of amino acids	C	A	X	Prenylation
CAAX	cysteine	any amino acid except alanine	serine, alanine, methionine, or glutamine	farneysylation
			leucine or isoleucine	geranylgeranylation
			phenylalanine	Either farnesylation or geranylgeranylation

Table 1.
Consensus sequences and prenylation.

is possible if X is phenylalanine. Thus "X" residue is largely responsible for determining which isoprenoid is attached to the protein target **Table 1** [17].

Farnesyltransferase (FTase), geranylgeranyltransferase type 1 (GGTase-I) Geranylgeranyltransferase type 2 (GGTase-II), and Geranylgeranyltransferase type 3 (GGTase-III) are the enzymes collectively referred to as the prenyltransferases. These prenyltransferases catalyze the addition of the prenyl groups to the proteins which is crucial for their activities **Table 2.**

FTase and GGTase-I are αβ heterodimeric proteins and share a common α-subunit (FNTA) and different β-subunits (FNTB and PGGT1B, respectively) [18, 19]. GGTase-II is also a heterodimer, having distinct α and β subunits (RABGGTA and RABGGTB, respectively). A third subunit denoted as Rab escort protein (REP) forms a very important functional part of GGTase-II [20]. The cytosolic fraction of all tissues including the brain contains all these three enzymes. The active sites of FTase and GGTase-I completely cover the isoprenoid and major parts of the C-terminus of the substrate protein and their three-dimensional structures exhibit high similarities. Recent reports point out that the FTase can prenylate cysteine residues efficiently at C-terminus position number five instead of the usual four [21]. This raises the possibility of the number of prenylated proteins to be larger than the one already envisaged. FTase is more selective compared to GGTase-I. While GGTase-I accepts many of the FTase substrates, the converse is generally not true with very few exceptions [22].

In addition to the CAAX, GGTase-II can also recognize CC, CXC, CCX, CCXX, and CCXXX at the C- terminus [23, 24]. Both cysteine residues within the sequences noted above are geranylgeranylated by GGTase-II.

The substrate specificities determined for GGTase-I and GGTase-II suggested that a third type of enzyme GGTase III may be found to catalyze the geranylgeranylation of the CXC terminating proteins [25, 26]. GGTase-III consists of prenyltransferase α

Prenylation reaction	Enzyme	Examples of prenylated substrates
Farnesylation	FTase	Ras proteins, lamin A, lamin B, CENP-E, CENP-F, transducin γ subunit
Geranylgeranylation	GGTase I	Rho proteins, Rac proteins, Ral proteins, Rap proteins, Gγ subunit
	GGTase II	Rab proteins
	GGTase III	Ykt6, FBXL2

Table 2.
Prenyl-modifying enzymes and their protein substrates.

subunit repeat-containing 1 (PTAR1) and the β subunit of RabGGTase. GGTase-III has been identified using a biotinylated geranylgeranyl analogue, the Golgi SNARE protein Ykt6, and ubiquitin ligase FBXL2 as a substrate. GGTase-III transfers a geranylgeranyl group to mono-farnesylated Ykt6, generating doubly prenylated Ykt6 [27].

Generally, the proteins after prenylation with few exceptions, get localized in the endoplasmic reticulum for further processing [28]. The two proteases namely Ras converting enzyme (Rce1) or Zinc metalloproteinase Ste24 (ZMPSte24) cleave the "AAX" residues from the CAAX-box on farnesylated proteins [29, 30].

2.1 Farnesylation

Zinc is used to coordinate the binding of farnesyl moiety to the thiol or sulfhydryl group of the cysteine in the CAAX box, thus pointing to the fact that FTase is a metalloenzyme. The FPP is bound by the FTase first, which is then followed by the binding of the protein. The release of the farnesylated protein occurs only after the farnesyl moiety is transferred to the protein. Farnesylation is a very essential modification in proteins to enable them to bind to membranes and thereby mediate signal transduction, both in normal as well as malignant cells [31].

Farnesylated proteins need a second signal for binding to the membrane since the farnesyl moiety is not sufficiently hydrophobic to stably anchor a protein to the membrane. The essential second signal is either in the form of one or two palmitate moieties attached to the cysteine upstream to the farnesylated cysteine residue or a polybasic K-Ras4B motif [32]. The movement of farnesylated K-Ras4B in the cytosol of the cell is regulated by the farnesyl electrostatic switches [33].

2.2 Geranylgeranylation

Unlike the farnesyl moiety the geranylgeranyl moiety is sufficiently hydrophobic. This makes the membrane binding of the modified protein more stable. A mechanism to extract the geranylgeranylated Rab and Rho proteins from the membrane using protein-protein interactions is observed, though the geranylgeranyl moiety is not removed from the modified protein all through its lifetime. The GTP- bound forms of Rab and Rho proteins are bounded to the soluble Guanine Dissociation inhibitor (GDI) proteins. The hydrophobic cavity of the GDI binds to at least one isoprenoid moiety. The inactivation of the activated Rab and Rho in the cytosol is preceded by their removal from the membrane as a result of isoprenoid sequestration [32].

3. Diversity of prenylated proteins and their functions

A large number of proteins have been identified which have undergone prenylation including several important intracellular proteins like heterotrimeric G protein subunits, nuclear lamins, Ras superfamily of small GTPases comprising of more than 150 known members [34]. This superfamily is divided into five subfamilies based on sequence and functional similarities: Ras, Rho, Rab, Ran, and Arf [35, 36]. During signal transduction, these proteins act as molecular switches cycling between "on" and "off" states. They regulate a variety of cellular functions like growth, differentiation and proliferation, vesicular trafficking by interacting with the downstream effectors. Each subfamily interacts with multiple effector proteins downstream. The activation of the downstream signaling pathway is dependent on its association with the cellular membrane which in turn relies on prenylation [37]. The prenylation of the small GTPases impacts its structure and function and in turn, affects the functions of the effector proteins downstream.

3.1 Nuclear lamins

The major architectural proteins of the animal cell nucleus are constituted by the lamins, which line inside the nuclear membrane and provide a platform for the binding proteins. They belong to the intermediate filament (IF) family of proteins and constitute an important part of the cytoskeleton. Based on the genomic structure and nucleotide sequence the nuclear lamins represent type V of the six types of IF superfamily. Based on the structural and protein features and expression patterns the nuclear lamins are divided into A and B types. Type A lamind are generally expressed in developmentally regulated temporal patterns but the Type B is expressed ubiquitously. Lamins contain a CAAX box at the C-terminus with few exceptions. These CAX boxes serve as sites of post-translational farnesylation. The isoprenylation of the cysteine is followed by the proteolytic cleavage of the AAX motif. Carboxy methylation of cysteine is the last of the three successive steps in the modification. After the cell division, the nuclear reassembly is mediated by the lamins. Type A lamins are distributed in the cytoplasm as they are easily solubilized whereas Type B lamins are closely associated with the nuclear membrane. The differential farnesylation of the lamin protein leads to their differences in membrane attachment during mitosis. During mitosis, the mature Type B lamins are anchored to the membrane by the farnesyl moiety and the Type A lamins are more soluble in the cytoplasm as the farnesyl moiety is removed [38].

3.2 Ras superfamily of small GTPase

The H-Ras, K-Ras, and N-Ras are the best-studied Ras isoforms owing to their oncogenic roles. They undergo the three-step modification namely prenylation, proteolysis, and carboxymethylation at the membrane of the endoplasmic reticulum. The important biological functions of Ras-like interaction with regulatory proteins, effector protein downstream including the signal transduction occur at the plasma membrane. Regulated by sophisticated mechanisms, the Ras proteins switch between inactive GDP-bound or active GTP-bound forms at the plasma membrane. In response to the stimulation by the extracellular signals through the cell surface receptors, the Ras proteins are switched on by guanine nucleotide exchange factors (GEFs), which catalyze the release of GDP and the binding of GTP. The Ras proteins are switched off by the GTPase-activating proteins (GAPs) which accelerate the intrinsic GTPase activity of Ras for hydrolysis of GTP. This Ras GDP/GTP switch is further regulated by guanine dissociation inhibitors (GDIs) and GDI-like proteins. A conformational change in Ras proteins is induced by the replacement of GDP for GTP which allows them to interact with their downstream effectors and execute their multiple signaling functions [39]. The selective interaction of K-Ras with membrane lipids leads to the formation of "nanoclusters" that determine it signaling output. The lipid specificity and signaling output of the K-Ras can be altered by subtle changes to its membrane anchor sequence or prenylation [39]. The importance of the prenylation in the functional regulation of Ras renders the isoprenoids and prenyltransferases as viable targets for therapeutic purposes.

4. Prenylated proteins and their functional regulation

The intracellular GTP-binding proteins are the largest family of prenylated proteins. They transduce extracellular signals into intracellular changes via downstream effectors. These proteins include the heterotrimeric G protein subunits and the small G protein superfamily. Despite the fact that both heterotrimeric and small G proteins have GDP/GTP-binding and GTPase activity only the small G proteins are commonly referred to as GTPases. The small G proteins are monomeric proteins and they are structurally and functionally distinct from the heterotrimeric G proteins. The heterotrimeric G proteins are activated by agonist-bound G protein-coupled receptors (GPCRs). GTP exchange on small G proteins is controlled by GEFs that catalyze the exchange of GDP for GTP. The functions of many of the small G proteins can be classified broadly as follows: regulation of gene expression by Ras subfamily; regulation of both cytoskeletal reorganization and gene expression by the Rho/Rac/Cdc42 proteins of the Rho family; regulation of intracellular vesicle trafficking the Rab and Sar1/Arf proteins, and regulation of nucleocytoplasmic transport during the G1, S and G2 phases of the cell cycle the Ran family members [40].

4.1 Guanine dissociation inhibitors (GDIs)

The pivotal regulators of small G protein function are the GDIs. The activation of the G proteins is prevented by Rho GDIs via three distinct mechanisms. The isoprenoid group of the GTPase is inserted into the hydrophobic pocket formed by the immunoglobulin-like β sandwich of GDI and thus shielded from the solvent. This enables the GDIs to maintain the Rho GTPases as soluble cytosolic proteins forming high-affinity complexes. The dissociation of GDP from Rho proteins is inhibited by GDIs, preventing GTPase activation by GEFs. Finally, the GDIs are able to interact with the GTP-bound form of the GTPase to prevent interactions with effector targets [41, 42].

4.2 Guanine nucleotide exchange factors (GEFs)

After dissociation from the GDIs, an extremely slow process catalyzed by the GEFs is the association of small G proteins with plasma membranes and exchange of GDP for GTP. The general mechanistic features of GEF-catalyzed GDP dissociation from Rho family proteins appear to be conserved for Ras family proteins and their GEFs. The GDP-GTP exchange reaction is thought to be the rate-limiting step in the GTP-binding/GTP hydrolytic cycle of GTPases and the GEFs have therefore are the key regulators of the GTPases.

4.3 GTPase-activating proteins (GAPs)

The action of the intrinsic GTpases converts GTP-bound forms of the GTPases to the inactive GDP-bound forms. Hydrolysis of GTP by intrinsic GTPases occurs slowly under normal circumstances, but a further group of regulatory proteins known as GAPs can accelerate this process. These proteins have important and widespread roles in the regulation of the small GTPases. GAPs are classified according to their GTPase subfamily (Ras-GAP, Rap-GAP, etc.) with sequence homology within subfamilies but not between families [43].

5. Prenylation and cancer

The prenylated proteins are ubiquitous in nature and have diverse roles in cellular biology. Detailed research on the oncogenic potential of prenylated proteins has been stimulated by the fact that mutations within the Ras family could be identified in as many as 10–15% of all human cancers [44, 45]. New insights into the role of farnesylated proteins in neoplasia have been provided by the use of farnesyl transferase knockout mouse by showing that farnesylated proteins apparently are not required for the initial malignant cellular transformation but are critical for tumor progression and maintenance. Of the GTPase regulatory proteins, a number of GEFs have also been associated with human malignancy. A gene rearrangement of GEF gene has also been identified in a patient with acute myelogenous leukemia. The drug discovery endeavors are directed towards the development of farnesyltransferase inhibitors (FTIs) as novel anticancer drugs owing to the central role of prenylated proteins in malignancy [46–48]. Many CAAX peptidomimetics, as well as FTIs and geranylgeranyltransferase inhibitors (GGTIs) have been developed that are highly potent and with low toxicity [49, 50]. The anchorage independent growth of cancer cells is inhibited by FTIs causing changes in cell cycle progression which further induce the apoptosis of cancer cells and inhibition of their attachment to extracellular substratum [51]. It has been shown that FTIs can also inhibit GGTase-II along with FTase, resulting in the blocking of post-translational modification of Rab proteins, leading to cell death [52]. This interference with endosomal trafficking is attributed as a novel action of FTIs. A number of prenylated proteins have been shown to be involved in the initiation, invasion, and progression of cancer in addition to the Ras proteins [53]. A search for important enzymes within the prenylation pathway as potential anticancer therapeutics is underway. Prenylated proteins also appear to play a role in both benign and malignant bone disease. The bisphosphonates are drugs that inhibit farnesyl diphosphate synthase resulting in effects on osteoclasts and tumor cells. Inhibition of protein prenylation and Ras signaling within osteoclasts leads to defects in intracellular vesicle transport that result in the imperfect formation of the tight-sealing zones or ruffled borders required for bone reabsorption [54]. Aside from their antiproliferative effects via inhibition of Ras, bisphosphonates also appear to have important proapoptotic effects that may be mediated by the accumulation of isopentenyl diphosphonate. Isopentanyl diphosphonate can be metabolized to Apppi (triphosphoric acid 1-adenosine-5'-yl ester 3-[3-methylbut-3-enyl] ester), an intracellular ATP analogue that can directly induce apoptosis [55]. To suppress the activity of the oncogenic Ras proteins to achieve the antitumor activity, the inhibition of prenylation and related pathways are being extensively investigated.

6. Conclusion

Prenylation is an irreversible post-translational modification during the lifetime of the protein. It is responsible for the correct cellular localization, activity, and protein–protein interactions of a number of signaling proteins. Over the years, the structural and mechanistic features of the prenyltransferase enzymes have been probed using a large number of isoprenoid analogs. The enzymology of prenyltransferases is well understood with extensive studies using these analogs. The malignant activity of oncogenic Ras proteins can be suppressed by using prenyltransferase inhibitors. But much remains to be learned concerning the roles of prenylated proteins in living cells, and this remains an intense area of current investigation. Site-specific modifications of proteins are a new possibility with prenylation. Overall, the challenges in the investigation of protein prenylation will keep this vibrant field of inquiry an active exciting endeavor in the near future.

Conflict of interest

The authors declare no conflict of interest.

List of Abbreviations

GPI	Glycosylphosphatidylinositol-linked proteins
FPP	Farnesyl pyrophosphate
GGPP	Geranylgeranyl pyrophosphate
FPS	Farnesyl pyrophosphate synthase
GGPPS	Geranylgeranyl diphosphate synthase
IPP	Isopentenyl pyrophosphate
IF	Intermediate Filaments
DMAPP	Dimethylallyl pyrophosphate
MVA	Mevalonate
MEP	Methylerythritol 4-phosphate pathway
FPS	Farnesyl pyrophosphate synthase
GGPPS	Geranylgeranyl diphosphate synthase
FTase	Farnesyltransferase
GGTase-I	Geranylgeranyltransferase type 1
GGTase-II	Geranylgeranyltransferase type 2
GGTase-III	Geranylgeranyltransferase type 3
FNTA	FarNesylTransferase, Alpha subunit
FNTB	FarNesylTransferase, Beta subunit
PGGT1B	Protein GeranylGeranylTransferase type I subunit Beta
RAB	Ras-associated binding
Ras	Rat sarcoma
RABGGTA	Rab Geranylgeranyltransferase Subunit Alpha
RABGGTB	Rab Geranylgeranyltransferase Subunit Beta
REP	Rab escort protein
PTAR1	Prenyltransferase α subunit repeat containing 1
Rce1	Ras converting enzyme
ZMPSte24	Zinc metalloproteinase Ste24
GEFs	Guanine nucleotide exchange factors
GDIs	Guanine dissociation inhibitors
GPCRs	G protein coupled receptors
GAPs	GTPase-activating proteins
FTIs	Farnesyltransferase inhibitors
GGTIs	Geranylgeranyltransferase inhibitors

Author details

Aravind Kamath and Kantharaju Kamanna*
Department of Chemistry, Peptide and Medicinal Chemistry Research Laboratory, Rani Channamma University, Belagavi, Karnataka, India

*Address all correspondence to: kk@rcub.ac.in

References

[1] Novelli G, D'Apice MR. Protein farnesylation and disease. Journal of Inherited Metabolic Disease. 2012;**35**(5):917-926. DOI: 10.1007/s10545-011-9445-y

[2] Ambrogelly A, Palioura S, Soll D. Natural expansion of the genetic code. Nature Chemical Biology. 2007;**3**:29-35

[3] Folch J, Ascoli I, Lees M, Meath JA, Lebaron FN. Preparation of lipide extracts from brain tissue. The Journal of Biological Chemistry. 1951;**191**(2):833-841

[4] Kamiya Y, Sakurai A, Tamura S, Takahashi N. Structure of rhodotorucine A, a novel lipopeptide, inducing mating tube formation in Rhodosporidium toruloides. Biochemical and Biophysical Research Communications. 1978;**83**(3):1077-1083

[5] Wolda SL, Glomset JA. Evidence for modification of lamin B by a product of mevalonic acid. The Journal of Biological Chemistry. 1988;**263**(13):5997-6000

[6] Farnsworth CC, Wolda SL, Gelb MH, Glomset JA. Human lamin B contains a farnesylated cysteine residue. The Journal of Biological Chemistry. 1989;**264**(34):20422-20429

[7] Sacchettini JC, Poulter CD. Creating isoprenoid diversity. Science. 1997;**277**(5333):1788-1789

[8] Houten SM, Frenkel J, Waterham HR. Isoprenoid biosynthesis in hereditary periodic fever syndromes and inflammation. Cellular and Molecular Life Sciences. 2003;**60**(6):1118-1134

[9] Kovacs WJ, Olivier LM, Krisans SK. Central role of peroxisomes in isoprenoid biosynthesis. Progress in Lipid Research. 2002;**41**(5):369-391

[10] Hogenboom S, Tuyp JJ, Espeel M, Koster J, Wanders RJ, Waterham HR. Mevalonate kinase is a cytosolic enzyme in humans. Journal of Cell Science. 2004;**117**(Pt 4):631-639

[11] Flesch G, Rohmer M. Prokaryotic hopanoids: The biosynthesis of the bacteriohopane skeleton. European Journal of Biochemistry. 1988;**175**: 405-411. DOI: 10.1111/j.1432-1033.1988.tb14210.x

[12] Rohmer M, Knani M, Simonin P, Sutter B, Sahm H. Isoprenoid biosynthesis in bacteria: A novel pathway for the early steps leading to isopentenyl diphosphate. Biochemical Journal. 1993;**295**:517-524

[13] Rodríguez-Concepción M, Boronat A. Elucidation of the methylerythritol phosphate pathway for isoprenoid biosynthesis in bacteria and plastids. A metabolic milestone achieved through genomics. Plant Physiology. 2002;**130**:1079-1089

[14] Eisenreich W, Bacher A, Arigoni D, Rohdich F. Biosynthesis of isoprenoids via the non-mevalonate pathway. Cellular and Molecular Life Sciences. 2004;**61**(12):1401-1426

[15] Rodríguez-Concepción M. The MEP pathway: A new target for the development of herbicides, antibiotics and antimalarial drugs. Current Pharmaceutical Design. 2004;**10**(19):2391-2400

[16] Jeong A, Suazo KF, Wood WG, Distefano MD, Li L. Isoprenoids and protein prenylation: Implications

in the pathogenesis and therapeutic intervention of Alzheimer's disease. Critical Reviews in Biochemistry and Molecular Biology. 2018;**53**(3):279-310. DOI: 10.1080/10409238.2018.1458070

[17] Zhang FL, Casey PJ. Protein prenylation: Molecular mechanisms and functional consequences. Annual Review of Biochemistry. 1996;**65**:241-269

[18] Casey PJ, Solski PA, Der CJ, Buss JE. p21ras is modified by a farnesyl isoprenoid. Proceedings of the National Academy of Sciences of the United States of America. 1989;**86**(21):8323-8327

[19] Casey PJ, Thissen JA, Moomaw JF. Enzymatic modification of proteins with a geranylgeranyl isoprenoid. Proceedings of the National Academy of Sciences of the United States of America. 1991;**88**(19):8631-8635

[20] Seabra MC, Goldstein JL, Südhof TC, Brown MS. Rab geranylgeranyl transferase. A multisubunit enzyme that prenylates GTP-binding proteins terminating in Cys-X-Cys or Cys-Cys. The Journal of Biological Chemistry. 1992;**267**(20):14497-14503

[21] Blanden MJ, Suazo KF, Hildebrandt ER, Hardgrove DS, Patel M, Saunders WP, et al. Efficient farnesylation of an extended C-terminal C(x) 3X sequence motif expands the scope of the prenylated proteome. The Journal of Biological Chemistry. 2018;**293**(8):2770-2785

[22] Krzysiak AJ, Aditya AV, Hougland JL, Fierke CA, Gibbs RA. Synthesis and screening of a CaaL peptide library versus FTase reveals a surprising number of substrates. Bioorganic and Medicinal Chemistry Letters. 2010;**20**(2):767-770

[23] Kinsella BT, Maltese WA. rab GTP-binding proteins with three different carboxyl-terminal cysteine motifs are modified in vivo by 20-carbon isoprenoids. The Journal of Biological Chemistry. 1992;**267**(6):3940-3945

[24] Pereira-Leal JB, Seabra MC. The mammalian Rab family of small GTPases: Definition of family and subfamily sequence motifs suggests a mechanism for functional specificity in the Ras superfamily. Journal of Molecular Biology. 2000;**301**(4):1077-1087

[25] Farnsworth CC, Kawata M, Yoshida Y, Takai Y, Gelb MH, Glomset JA. C terminus of the small GTP-binding protein smg p25A contains two geranylgeranylated cysteine residues and a methyl ester. Proceedings of the National Academy of Sciences of the United States of America. 1991;**88**(14):6196-6200

[26] Khosravi-Far R, Lutz RJ, Cox AD, Conroy L, Bourne JR, Sinensky M, et al. Isoprenoid modification of rab proteins terminating in CC or CXC motifs. Proceedings of the National Academy of Sciences of the United States of America. 1991;**88**(14):6264-6268

[27] Shirakawa R, Goto-Ito S, Goto K, Wakayama S, Kubo H, Sakata N, et al. A SNARE geranylgeranyltransferase essential for the organization of the Golgi apparatus. EMBO Journal. 2020;**39**(8):e104120

[28] Hildebrandt ER, Cheng M, Zhao P, Kim JH, Wells L, Schmidt WK. A shunt pathway limits the CaaX processing of Hsp40 Ydj1p and regulates Ydj1p-dependent phenotypes. eLife. 2016;**15**(5):e15899

[29] Boyartchuk VL, Ashby MN, Rine J. Modulation of Ras and a-factor function by carboxyl-terminal proteolysis. Science. 1997;**275**(5307):1796-1800

[30] Chen Y, Ma YT, Rando RR. Solubilization, partial purification, and

affinity labeling of the membrane-bound isoprenylated protein endoprotease. Biochemistry. 1996;**35**(10):3227-3237

[31] Prior IA, Hancock JF. Ras trafficking, localization and compartmentalized signalling. Seminars in Cell and Developmental Biology. 2012;**23**(2): 145-153

[32] Resh MD. Chapter 13—Lipid modification of proteins. In: Ridgway ND, McLeod RS, editors. Biochemistry of Lipids, Lipoproteins and Membranes. 7th ed. Amsterdam: Elsevier Science; 2021. pp. 429-456. Paperback. ISBN: 9780128240489 eBook ISBN: 9780323983280

[33] Bivona TG, Quatela SE, Bodemann BO, Ahearn IM, Soskis MJ, Mor A, et al. PKC regulates a farnesyl-electrostatic switch on K-Ras that promotes its association with Bcl-XL on mitochondria and induces apoptosis. Molecular Cell. 2006;**21**(4):481-493

[34] McTaggart SJ. Isoprenylated proteins. Cellular and Molecular Life Sciences. 2006;**63**(3):255-267

[35] Colicelli J. Human RAS superfamily proteins and related GTPases. Science STKE. 2004;**250**:RE13

[36] Wennerberg K, Rossman KL, Der CJ. The Ras superfamily at a glance. Journal of Cell Science. 2005;**118**(Pt 5):843-846

[37] Lane KT, Beese LS. Thematic review series: Lipid posttranslational modifications. Structural biology of protein farnesyltransferase and geranylgeranyltransferase type I. Journal of Lipid Research. 2006;**47**(4):681-699

[38] Sinensky M, Fantle K, Trujillo M, McLain T, Kupfer A, Dalton M. The processing pathway of prelamin A. Journal of Cell Science. 1994;**107**(Pt 1):61-67

[39] Cherfils J, Zeghouf M. Regulation of small GTPases by GEFs, GAPs, and GDIs. Physiological Reviews. 2013;**93**(1):269-309

[40] Zhou Y, Prakash P, Liang H, Cho KJ, Gorfe AA, Hancock JF. Lipid-sorting specificity encoded in K-RAS membrane anchor regulates signal output. Cell. 2017;**168**(1-2):239-251

[41] Takai Y, Sasaki T, Matozaki T. Small GTP-binding proteins. Physiological Reviews. 2001;**81**(1):153-208

[42] DerMardirossian C, Bokoch GM. GDIs: Central regulatory molecules in Rho GTPase activation. Trends in Cell Biology. 2005;**15**(7):356-363

[43] Hoffman GR, Nassar N, Cerione RA. Structure of the Rho family GTP-binding protein Cdc42 in complex with the multifunctional regulator RhoGDI. Cell. 2000;**100**(3):345-356

[44] Donovan S, Shannon KM, Bollag G. GTPase activating proteins: Critical regulators of intracellular signaling. Biochimica et Biophysica Acta. 2002;**1602**(1):23-45

[45] Downward J. Targeting RAS signalling pathways in cancer therapy. Nature Reviews. Cancer. 2003;**3**(1):11-22

[46] Sahai E, Marshall CJ. RHO-GTPases and cancer. Nature Reviews. Cancer. 2002;**2**(2):133-142

[47] Gibbs JB, Oliff A, Kohl NE. Farnesyltransferase inhibitors: RAS research yields a potential cancer therapeutic. Cell. 1994;**77**(2):175-178

[48] Sebti SM, Der CJ. Opinion: Searching for the elusive targets of farnesyltransferase inhibitors. Nature Reviews. Cancer. 2003;**3**(12):945-951

[49] Doll RJ, Kirschmeier P, Bishop WR. Farnesyltransferase inhibitors as anticancer agents: Critical crossroads. Current Opinion in Drug Discovery & Development. 2004;**7**(4):478-486

[50] Mazieres J, Pradines A, Favre G. Perspectives on farnesyl transferase inhibitors in cancer therapy. Cancer Letters. 2004;**206**(2):159-167

[51] Tamanoi F, Gau CL, Jiang C, Edamatsu H, Kato-Stankiewicz J. Protein farnesylation in mammalian cells: Effects of farnesyltransferase inhibitors on cancer cells. Cellular and Molecular Life Sciences. 2001;**58**(11):1636-1649

[52] Lackner MR, Kindt RM, Carroll PM, Brown K, Cancilla MR, Chen C, et al. Chemical genetics identifies Rab geranylgeranyl transferase as an apoptotic target of farnesyl transferase inhibitors. Cancer Cell. 2005;**7**(4):325-336

[53] Winter-Vann AM, Casey PJ. Post-prenylation-processing enzymes as new targets in oncogenesis. Nature Reviews. Cancer. 2005;**5**(5):405-412

[54] Alakangas A, Selander K, Mulari M, Halleen J, Lehenkari P, Mönkkönen J, et al. Alendronate disturbs vesicular trafficking in osteoclasts. Calcified Tissue International. 2002;**70**(1):40-47

[55] Green JR. Bisphosphonates: Preclinical review. The Oncologist. 2004;**9**(Suppl. 4):3-13

Chapter 4

Protein Prenylation and Their Applications

Khemchand R. Surana, Ritesh B. Pawar, Ritesh A. Khairnar and Sunil K. Mahajan

Abstract

Prenylation is a universal covalent post-translational modification found in all eukaryotic cells, comprising attachment of either a farnesyl or a geranylgeranyl isoprenoid. Prenyl group is important for protein-protein binding through specialized prenyl-binding domains. Farnesylation and geranyl geranylation are very important in C-terminal anchoring of proteins to the cell membrane. These post-translational modification are most often catalyzed by either protein farnesyl transferase (FTase) or protein geranyl geranyl transferase-I (GGTase-I). These enzymes typically recognize a CaaX motif, where "C" is the cysteine to be prenylated and the remainder of the motif leads to recognition by FTase and/or GGTase-I. Prenylation plays vital role in diversification of natural products flavonoids, coumarins, and isoflavonoids. Many prenylated compounds have been identified as active components in medicinal plants with biological activities, such as anti-cancer, anti-spasmodic, anti-bacterial, anti-fungal, anti-inflammatory, and anti-androgen activity. Due to their beneficial effects on diseases, prenylated compounds are of particular interest as lead compounds for producing drugs and functional foods. In this chapter, we concise the prenylation reactions of aromatic compounds such as indole, ketones, and aldehydes that may results to lead molecules discovery. Prenylation reactions are applied on azoles, anilines, thioles, indole, α-carbonyl bromides, and aryl bromide. There are several drugs that are obtained from prenylation, i.e. (-)-17-hydroxy-citrinalin, (+)-stephacidin, prenylated. In this text there is no referencing, it is a chemical name, so keep as it is.

Keywords: prenylation reaction, prenylating agent, farnesyl, geranyl, natural products

1. Introduction

Prenylation is class of modification of molecules involving irreversible covalent bonding of isoprenoid unit to chemical compound or protein. Prenylation is known as lipidation or isoprenylation (**Figure 1**) [1].

Prenylation of protein is required to make protein fully functional. In a three-step process, farnesyl (15-carbon) or geranylgeranyl (20-carbon) group is transferred to protein, which occurs on carboxyl terminal of cysteine residues of protein. The RAM mutant, a *Saccharomyces cerevisiae* mutant, was used to characterize the location of

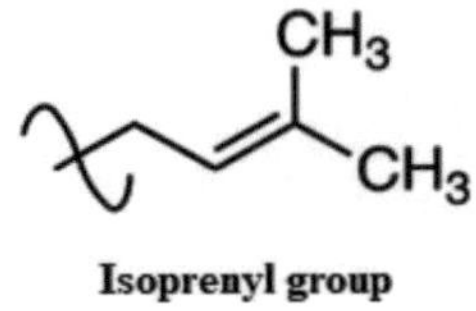

Figure 1.
Structure of isoprenyl group.

prenylation in proteins changed by these substituents. This mutant was found to be defective in the post-translational processing of both the ras gene product and the yeast a-mating factor. A common carboxy terminal region was discovered when the predicted amino acid sequences of these two proteins, other Ras proteins and mating factors, and the known prenylated proteins lamin B and prelamin A were compared. A CAAX box is a carboxy terminal amino acid sequence in which C is cysteine, A is an aliphatic amino acid, and X is the carboxy terminal amino acid. The three AAX amino acids were removed from the structure of yeast mating hormone a-factor, which was discovered to be farnesylated and carboxymethylated at the carboxy terminal cysteine ("). This discovery anticipated that a carboxy terminal cysteine in mammalian ras protein would be prenylated with farnesyl. The modifications of proteins by prenylation increase lipophilicity of prenylated proteins for efficient anchoring on plasma membranes or organellar membranes. Prenylated proteins plays role in a number of signaling and regulatory pathways that are responsible for basic cell operations [1, 2]. The first prenylated polypeptide to be discovered in late 1970 and early 1980 which is the mating factor from the fungus Rhodosporidium toruloides which is an undecapeptide containing a C-terminal S-farnesyl-cysteine methyl ester. These mating factors are found as farnesyl group attached to cysteine residue of short peptide [3, 4]. Farnesyltransferase and geranylgeranyl transferase type 1 (GGTase-I) are enzymes that catalyze the attachment of a single farnesyl (15 carbon) or geranylgeranyl (20 carbon) isoprenoid group to proteins. The addition of two geranylgeranyl groups to two cysteine residues in sequences such as CXC, CCXX near the C-terminus of Rab proteins is catalyzed by geranylgeranyl transferase type 2 (GGTase-II or Rab geranylgeranyl transferase) (**Figures 2** and **3**) [5, 6].

Natural chemicals can be prenylated to add structural variety, change biological activity, and improve medicinal potential. Prenylated natural products are a large class of bioactive molecules with demonstrated medicinal properties such as

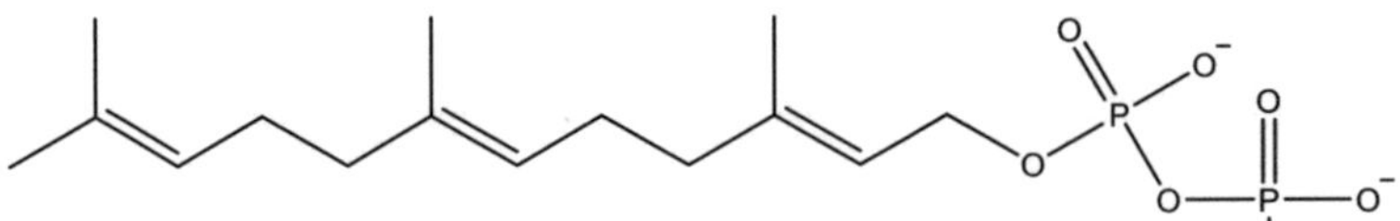

1

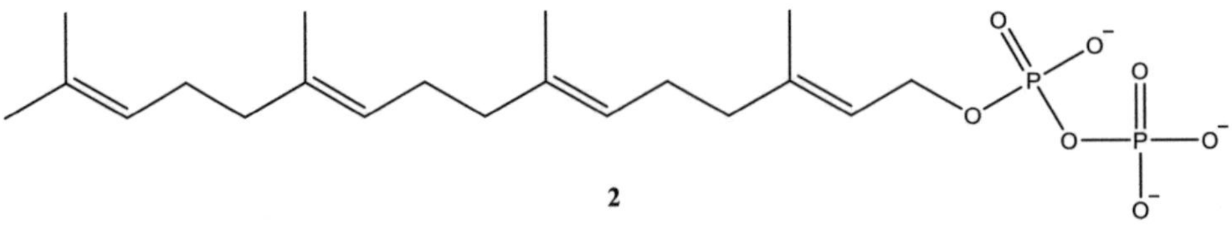

2

Figure 2.
Structures of 1 (farnesyl diphosphate, FPP) and 2 (geranylgeranyl diphosphate, GGPP).

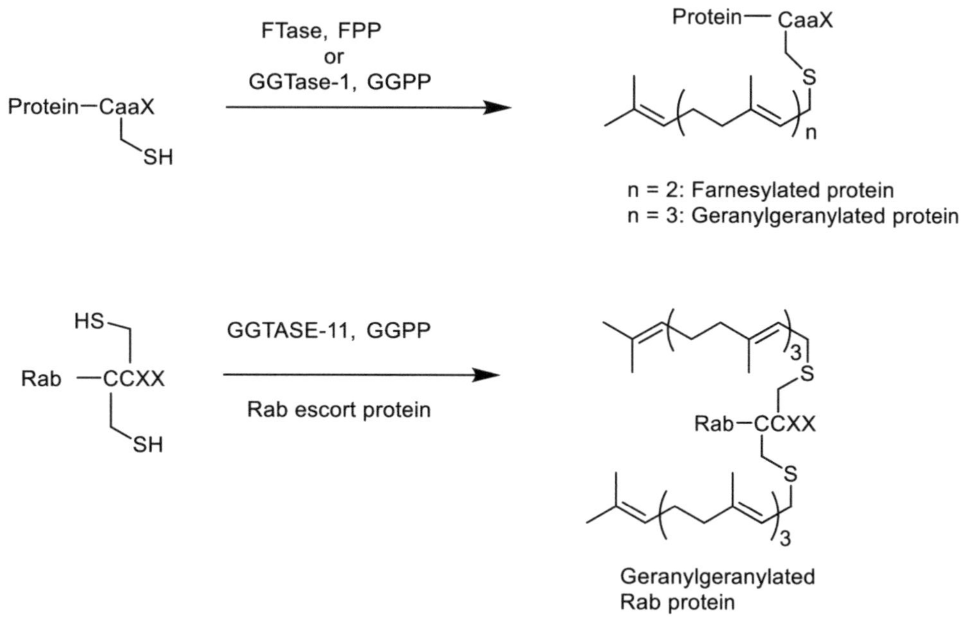

Figure 3.
Reactions catalyzed by prenyl transferase enzymes.

anti-emetics, anticonvulsants, antidepressants, and analgesics. Examples include prenyl-flavonoids, prenyl-stilbenoids, and cannabinoids. Prenyl transfer of natural products catalyzed by enzymes has high regio- and stereo-specificity, but it requires expensive isoprenyl pyrophosphate substrates [4]. In prenylated natural products, the carbon lengths of the prenyl side chains differ. Based on the size of the carbon, four different forms of prenyl side chains have been identified: C5 (isopentenyl), C10 (geranyl), C15 (farnesyl), and C20 (isopentenyl) (geranylgeranyl) [5]. Prenyltransferases catalyze the fundamental isoprenoid chain-elongation reaction, resulting in prenyl diphosphate with a variety of chain lengths and stereochemistries. All of the compounds are made up of linear isoprenoid diphosphates, which are made up of isopentenyl diphosphate (IPP) and dimethylallyl diphosphate, two isomeric 5-carbon unit intermediates (DMAPP). According to the specificities of particular prenyltransferases, the reactions are regulated to run continuously and to cease exactly at certain chain lengths [7].

2. History behind protein prenylation

Rhodotorucine A, a mating factor peptide of the yeast Rhodosporidium toruloides, was found in 1979 in Japan as the first example of prenylation of a carboxy-terminal cysteine residue. A S-farnesyl cysteine was discovered near the carboxy-terminus of the mature form. Over the next five years, more fungal peptidyls will be discovered. sex hormones covalently modified by a farnesyl group on a C-terminal cysteine were made public This modification's function was unknown, but it was stoichiometric and stable, indicating that it is a key component of the function. a list of mate-inducing variables Prenylation in mammals was discovered independently for the first time. Compactin, a cholesterol production inhibitor was discovered to have a negative effect on cells in 1984 [1].

Compactin inhibits 3-hydroxy-3-methylglutaryl-CoA-reductase (HMG-CoA reductase), an enzyme that produces mevalonic acid, a key step in the isoprenoid pathway. Compactin treatment of cultured cells resulted in cell cycle arrest and morphological changes. The fact that these effects could be reversed by giving the cells mevalonate but not cholesterol, dolichol, ubiquinone, or isopentenyl adenine, the principal products of isoprenoid biosynthesis, demonstrated that one or more non-sterol isoprenoids play a significant role in cell cycle and shape. Covalent incorporation of an isoprenoid derivative into cellular proteins was seen following the destiny of radiolabeled mevalonate in cultured 3T3 fibroblasts with their mevalonate production suppressed by lovastatin [3].

Isoprenylation was shown to be a common process in mammalian cells, including proteins from several compartments such as the nuclear envelope, plasma membrane, and cytoplasm. In 1986, a novel gene from the yeast Saccharomyces cerevisiae was discovered that was necessary for posttranslational modification of both the RAS proteins and an a-mating factor termed RAM. Except for the C-terminus, which consists of a cysteine followed by two residues with aliphatic side chains and an ultimate C-terminal amino acid residue, there was no evident sequence similarity between the yeast RAS proteins and the precursor polypeptide for a-factor. This so-called CaaX motif was discovered in a number of other proteins as well, although at the time, it was thought to communicate palmitoyl group alteration. New eukaryotic protein carboxyl methylation processes, such as modification of the subunit of cGMP phosphodiesterase and nuclear lamin B, were identified in 1985. These responses did not fall into any recognized activity category. The discovery of the CaaX-motif in the aforementioned proteins, as well as the carboxymethylation of several fungal mating factors at the C-terminus, led to the notion that the CaaX sequence signal related prenylation, proteolytic cleavage, and methylation.

The discovery in 1988 that the yeast a-factor contains a C-terminal farnesylcysteine methyl ester and the discovery in 1989 that all mammalian Ras proteins are isoprenylated on the conserved cysteine residue supported this hypothesis. Upstream cysteine residues in a subset of Ras proteins could be assigned to palmitoylation, which was previously thought to modify this amino acid. At the same time, lamin B was discovered to be the first isoprenylated protein. In 1990, trans, trans-farnesyl, and all-trans-geranylgeranyl were found as the primary isoprenoids bound to mammalian protein. The discovery that nuclear lamin B, mammalian Ras, and yeast RAS are farnesylated was followed by the discovery that a -subunit of a mammalian heterotrimeric G-protein is geranylgeranylated. The finding that Ras farnesylation is essential for oncogenic forms to convert cells spurred research efforts in this sector, resulting to the identification of the farnesyltransferase enzyme and the creation of particular protein prenylation inhibitors for cancer treatment [8].

3. Classes of prenylated proteins

The growing realization that the presence of a CAAX motif at a protein's carboxyl terminus indicated that it was a candidate for prenylation sparked a flurry of effort aimed at detecting the prenylation status of such proteins. Initial searches of protein sequence databases yielded 40 such possibilities; unexpectedly, the majority of these proteins were members of the GTP-binding protein class (sometimes known as "small G proteins"), which is linked to the ras proteins. Furthermore, the majority of the remaining possibilities (e.g., the subunits of heterotrimeric G proteins, cGMP phosphodiesterase) were recognized to have a role in cellular signaling. Almost all of the potential proteins that have been thoroughly examined have been discovered to be prenylated. Only the 15-carbon farnesyl and 20-carbon geranylgeranyl groups have been shown to alter proteins so far, with geranylgeranyl being the most often attached isoprenoid. The COOH-terminal amino acid ("X") of the CAAX box is now known to dictate which isoprenoid is linked to a candidate protein, which is of great interest to researchers investigating the enzymology of the prenylation processes (see below). The protein includes the farnesyl isoprenoid if this residue is a serine, methionine, or glutamine, whereas a leucine at that location drives geranylgeranyl addition. On proteins with the CAAX-motif, prenylation is not the sole posttranslational modification. The three COOH-terminal amino acids (the "AAX") are missing from the mature versions of these prenylated proteins. These three amino acids are removed by a cellular peptidase, leaving the prenylated cysteine as the COOH-terminal residue. Furthermore, a significant percentage of prenylated proteins have the carboxyl group of this cysteine residue methylated in all situations where the prenylated proteins have been thoroughly investigated. The end outcome of these three seemingly unrelated processing processes is a mature protein with a highly hydrophobic COOH-terminus, significantly improving the protein's inherent hydrophobic characteristics. It has recently been discovered that prenylation is not confined to proteins with the COOH-terminal C U X - motif, but also occurs on the rabIYPT1 family of GTP-binding proteins. The COOH-terminal sequences of the majority of these proteins finish in Cys-Cys or two cysteines separated by another amino acid (Cys-X-Cys). The rab/YPTl proteins appear to have the ability to reversibly connect with membranes and have been involved in the modulation of intracellular membrane trafficking pathways. Furthermore, both cysteine residues are prenylated with the geranylgeranyl isoprenoid in many (if not all) of these proteins. No proteolytic processing is necessary to expose the carboxylate of the prenylated cysteine residue for methylation because these alterations occur at the extreme COOH-terminus, although this methylation of the carboxyl group has been proven to occur [9].

4. Isoprenoid analogs

To examine various aspects of the prenylation reaction and prenyltransferases, a significant number of isoprenoid analogues with varied functions have been produced. Photo affinity probes like compounds 6 through 4 were employed extensively to investigate the structural properties of yeast and mammalian forms of FTase and GGTase-I before the crystal structure of FTase was discovered. These analogues revealed the role of the subunit of FTase and GGTase-I in isoprenoid substrate recognition and binding, as well as changes in active site layout between mammalian and yeast FTases. Similar tests using GGTase-II were later carried out, leading to the discovery of proteins with which Rab5 interacts via the isoprenoid group. While three was recognized as a substrate by FTase, four, which included one more isoprenoid unit, was found to be a powerful inhibitor of yeast FTase. A novel photoactive isoprenoid probe with a diazirine group was recently published, with a size that is closer to that of FPP. In cross-linking investigations of Icmt, peptides containing that photoactive isoprenoid were employed. Phosphonate 6 and related analogues 66 have been particularly effective in crystallographic investigations that have demonstrated that the isoprenoid binds in an extended conformation and that isoprenoid binding causes many active site residues to rearrange when compared to the unliganded

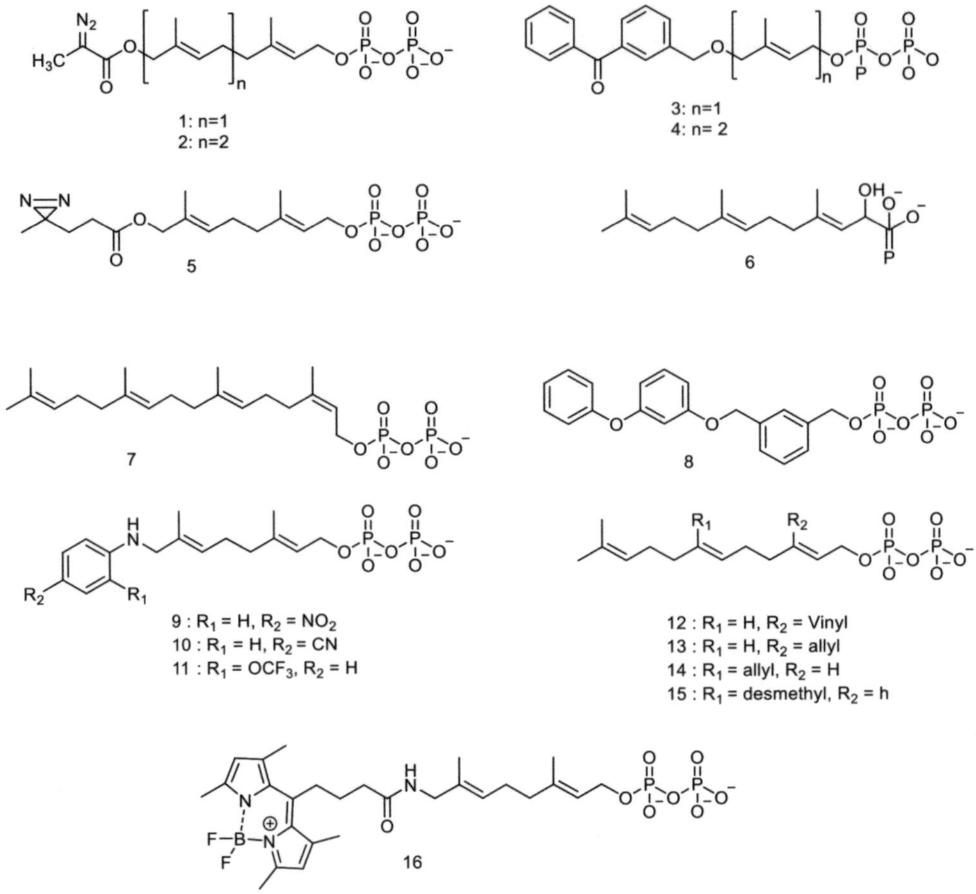

Figure 4.
Structures of isoprenoid analogs used to study structure, mechanism and isoprenoid substrate specificity of FTase and GGTase-I.

enzyme. Isoprenoid substrate specificity of prenyltransferase enzymes has been investigated recently using isoprenoid analogues. Gibbs and colleagues created a variety of all-trans FPP and GGPP geometric isomers. They discovered that whereas most of the analogues were substrates for mammalian FTase, they were not accepted as substrates by mammalian GGTase-I. Molecule 7 inhibited GGTase-I, showing a more rigorous selectivity for the enzyme's isoprenoid substrates. The binding site of FTase is very plastic, according to Spielmann and colleagues. Analog 8, for example, was an effective FTase substrate because all isoprene units were substituted with aryl groups. The anilinogeranyl-based isoprenoid analogues 9 and 10. The downstream enzymes Rce1 and Icmt processed proteins changed with these analogues, but the resulting modified proteins were not physiologically active, showing the necessity of enhanced hydrophobicity during prenylation. They also discovered that some anilinogeranyl-based analogues, such as 11, were substrates for FTase when a peptide-based on K-Ras C-terminal sequence, dansyl-GCVIM, was used, but that they were potent inhibitors of the enzyme when dansyl-GCVLS (IC50 of 16 was 3.0 nM), a sequence based on H-Ras C-terminal sequence, was used (**Figure 4**) [10].

Gibbs and colleagues synthesized a significant number of analogues with alterations at the 3- or 7-positions of FPP. They concluded that even minor modifications in the functionality incorporated at these points can result in significant and surprising differences in incorporation efficiency. For example, they discovered that the 3-vinyl analogue 12, which is a slow FTase substrate in cells, and the 3-allyl analogue, which is an FTase inhibitor, is both slow FTase substrates. During the screening of analogues against an eight-sequence CaaX library, the 7-allyl analogue 14, could only farnesylate

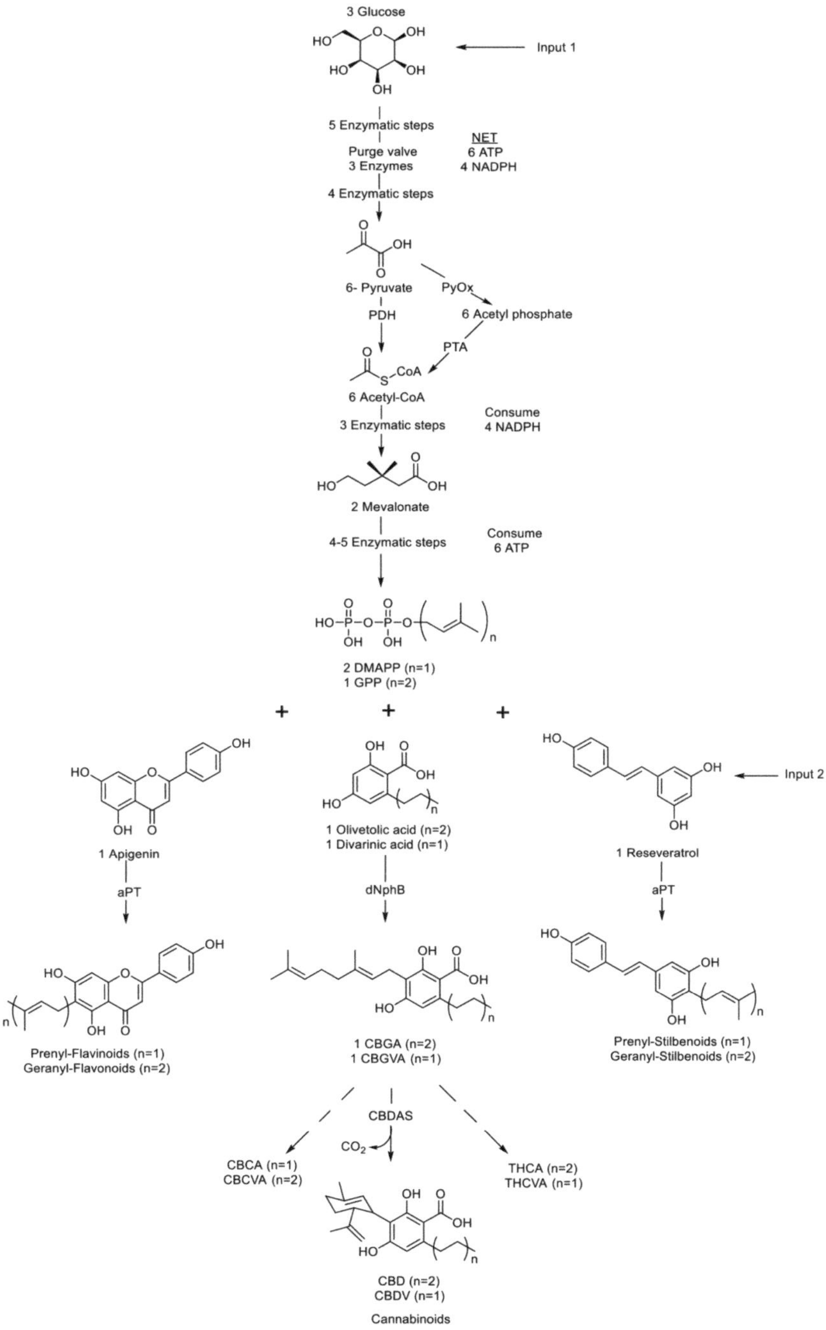

Figure 5.
The synthetic biochemistry platform for the production of prenyl natural products.

the CVIM sequence, whereas 15 was an exceptionally efficient substrate for practically all of the sequences in their library. These minor differences are likely due to the fact that when bound to FTase, the protein and isoprenoid substrates interact over a large surface area thus, small changes in one of the substrate structures necessitate compensatory changes in the other substrate to achieve optimal complementarity (**Figure 5**) [11, 12].

First, glucose is broken down into pyruvate through a glycolysis pathway modified to regulate NADPH levels (12 enzymatic steps). Then, either PDH or the PDH bypass converts pyruvate into acetyl CoA. Acetyl-CoA is converted into GPP via the mevalonate pathway (eight enzymatic steps). By varying the aromatic prenyltransferase (aPT) and aromatic substrate we can produce various prenyl-flavonoids and prenyl-stilbenoids using the same central pathway. We developed the prenyltransferase NphB (dNphB) variants to produce CBGA or CBGVA. CBGA is converted to cannabidiol and CBGVA is converted to cannabidivarin acid via cannabidiolic acid synthase (CBDAS). It is possible to produce other cannabinoids by using different cannabinoid synthases (THCAS and CBCAS) [13].

5. Mechanism of protein prenylation

5.1 Binding mechanism

There is currently a substantial body of knowledge about the role of the active site Zn^{2+} in PFTase. Mutagenesis of the protein-derived Zn^{2+} ligands resulted in significantly reduced Zn^{2+} binding and activity [14]. The replacement of Co^{2+} for Zn^{2+} provided more direct proof of Zn^{2+} in catalysis. Comparisons of the Co^{2+} –PFTase absorbance spectrum to that of model compounds revealed the presence of one thiolate ligand, possibly Cys299β, in the metal's inner coordination sphere. When peptide substrate was added, the spectrum resembled that of two thiolate ligands. This means that the cysteine in the peptide substrate is directly thiolate-coordinated to the metal. Cysteine ligation was later crystallographically verified with a Zn^{2+}–sulfur bond length of 2.5.52. Although the spectra of the Co^{2+} –PFTase–product complex differed from that of the ternary complex containing substrates, the Co^{2+} –PFTase–product complex showed more sulfur–cobalt coordination than the free enzyme. The farnesylated peptide interpreted the UV spectrum in terms of thioether coordination to Co^{2+}. Co^{2+}–PFTase has similar activity to the conventional Zn^{2+} enzyme. Single turnover tests utilizing absorbance stopped-flow spectrophotometry produced a two-exponential trace whose rates and extinction coefficient changes corresponded to the development of a Co^{2+} –PFTase–FPP–peptide ternary complex and subsequent conversion to the Co^{2+} –PFTase–product complex. On a time scale consistent with catalysis, the creation and disappearance of Co^{2+} –thiolate species were established [15].

5.2 Chemical mechanism

The molecular mechanism of the prenyltransferase reaction is currently being researched, and a better knowledge of the reaction's transition state should aid in the development of inhibitors for these enzymes. The diphosphate moiety in the prenyl substrate is displaced by the sulfur in the peptide/protein substrate in this reaction. It could happen through an associative mechanism in which the peptide thiolate attacks the C1 site of FPP directly, causing the diphosphate to be displaced (**Figure 6**).

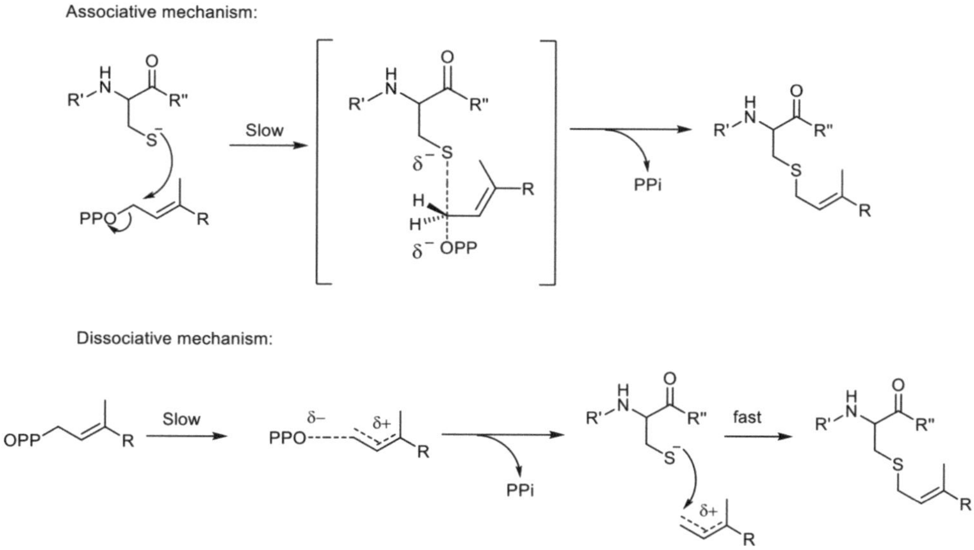

Figure 6.
Possible reaction mechanisms for protein prenyltransferases.

The reaction would continue through a single transition state in this situation. An associative mechanism can take a variety of forms, ranging from an early transition state with little positive charge growth in the allylic moiety of the prenyl substrate to a late transition state with significant positive charge development. The nucleophilicity of the substrate thiolate provides much of the "driving power" for an associative mechanism with an early transition state. The nucleophilicity of the thiolate and the capacity of the isoprene moiety to stabilize the growing positive charge are critical for a mechanism with a later transition state. An allylic carbocationic intermediate would be used in a dissociative process (**Figure 6**). The production of the pyrophosphate anion-allylic carbocation ion pair in the active site would result in significant charge separation. There would be two distinct transition states in this reaction. The stabilization of the allylic carbocation by resonance provides a significant portion of the "driving power" in this mechanism. L-b-farnesylaminoalanine (faA), a transition state analogue, was designed to mimic the charge and shape of a putative alkylated thiol intermediate formed by adding the prenyl group to a protonated sulfur [16]. The faA (IC50 = 51 mM) and tetrapeptide faA-VIA (IC50 = 14 mM) derivatives showed modest inhibitory activity. The IC50 of faA-VIA was reduced to 370 nM in the presence of inorganic pyrophosphate. This class of books has minimal binding. The fact that the thiolate form of cysteine is the species that assaults FPP is easily explained by subsequent research. Several fluorine derivatives of FPP were used to study the evolution of positive charge in the transition state. As the quantity of fluorines in the C3 methyl grows, a cationic intermediate becomes increasingly unstable. This approach produced an excellent Hammett plot for the rates of solvolysis Ksolv (a dissociative reaction) vs the rates of prenyl transfer in the reaction catalyzed by FPP synthase where the nucleophile is a carbon-carbon double bond, with a slope of 1 [17]. The enzyme-catalyzed reaction's response to substitution was thus very similar to a dissociative solvolysis reaction involving allylic carbocationic intermediates. Increased fluorine substitution revealed a high Hammett connection with the solvolysis hypothesis in the instance of PFTase. The slope, on the other hand, was significantly smaller than unity. Hammett plots of kcat vs the rate of associative displacement between allylic reactants and azide showed a considerably better correlation. The most obvious explanation of these findings is that the PFTase-catalyzed reaction's transition state is electrophilic, but the mechanism is associative with a late transition state with significant electrophilic character. The difference between PFTase and FPP synthase could be due to the thiolate prenyl acceptor's higher nucleophilicity when compared to a carbon-carbon double bond.

Weller and Distefano have mentioned measurement of the secondary kinetic isotope effect for prenylation of dansyl- GCVIA with [1-2H2] FPP [9]. They discovered kH/kD 1. If fully expressed, a dissociative mechanism should have a kH/kD of 1.2, and an associative mechanism with an early transition state should have a small inverse impact. The partial suppression of a kinetic isotope effect is predicted by commitment factors estimated from yeast PFTase rate constants. The lack of a detectable isotope effect supports an associative process, albeit it is unclear whether the experiments were done precisely enough to detect a minor isotope effect. R- and S-[1-2H] FPP were used to investigate the stereochemistry of the PFTase reaction [18]. The reaction preceded with inversion of configuration at C1 of FPP, according to the 1H NMR spectra of the products. While the stereochemical conclusion is compatible with an associative reaction, it could simply reflect the substrate's orientation in the active site prior to catalysis [19] Inversion of configuration has also been found for FPP synthase and dimethylallyl tryptophan synthase, both of which have strong evidence for a dissociative prenyl transfer mechanism. As a result, the stereochemical findings are compatible with either process. There is no precedent for an associative enzyme-catalyzed prenylation process other than PFTase. Dissociative alkylations appear to be catalyzed by FPP synthase and dimethylallyl tryptophan synthase, which prenylate weakly nucleophilic carbon-carbon double bonds and aromatic rings, respectively. Two copies of the highly conserved DDXXD or DDXXXXD motifs are found in FPP synthase and closely related chain elongation prenyltransferases [20].

The magnesium complexes of the allylic and homoallylic diphosphate substrates are recognized by these patterns. The aspartate-rich motif in FPP synthase has been subjected to site-directed mutagenesis, which suggests that these residues are critical for both catalysis and substrate binding. Although it is possible that a distinct structural pattern evolved for a dissociative alkylation, the absence of related aspartate-rich motifs in the active site of protein prenyltransferases could indicate that the mechanism for sulfur alkylation is different. A shift from the dissociative process reported in FPP synthase to an associative reaction for protein prenylation is consistent with a cysteine thiolate's high nucleophilicity [21].

6. Reagents for prenylation reaction and their applications

Prenylating reagent plays a significant function in bioactive and synthetic substances because it has a high binding affinity for proteins and increases membrane permeability, enhancing bioactivity and bioavailability of prenylated drugs. Through a specific prenyl-binding domain, the prenyl group plays an important role in protein-protein binding [22, 23].

6.1 Prenyl ester

Prenyl ester is used for the synthesis of prenyl alcohol. Prenyl ester is synthesis from alkanoic acid and isoprene. When alkanoic acid (it is a mixture of carboxylic

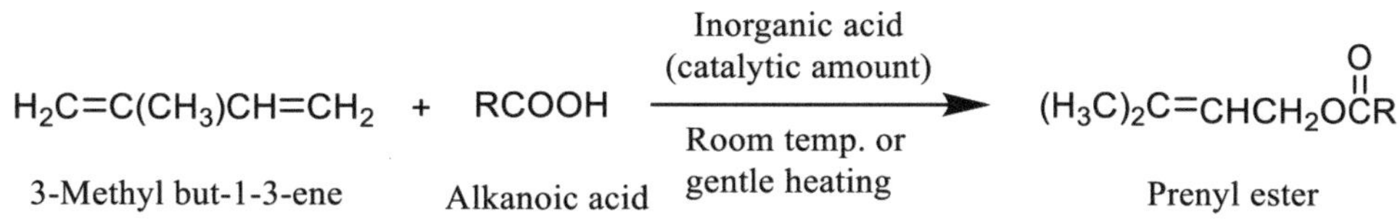

Figure 7.
Preparation of prenyl ester.

acid RCO_2H) Where R is a (C1-C2) alkyl group reacted with 3-Methyl but-1-3-ene (isoprene) in the presence of an inorganic acid catalyst to a mixture at room temperature or applying gentle heat to the reaction mixture will formation of Prenyl ester (**Figure 7**) [24].

When strong acidic condition giving to the reaction then reaction will occur however it will yield prenyl ester (Ex. Sulfuric acid or p-toluene sulfonic acid).

6.2 Prenyl acetate

Prenyl acetate is a derivative of prenyl ester. Prenyl ester is insoluble in water and soluble in heptanes and octane. Prenyl acetate is synthesized from alkanoic acid and isoprene.

When alkanoic acid (it is a mixture of carboxylic acid RCO_2H) Where R is a (C1-C2) alkyl group preferably acetic acid in molar excess is used for alkanoic acid, reacted with 2-methyl-1, 3- butadiene (isoprene) in the presence of phosphoric acid as the catalyst and reaction occurs at room temperature heating is not essential. However, absent heating, the reaction is slow.

Hence, gentle heating is giving to the reaction (about 40°C to about 100°C) of the reaction mixture under pressure (due to the volatility of isoprene). Then yield is form Prenyl Acetate (**Figure 8**) [25].

Formation of prenyl alcohol is one of the applications of Prenyl acetate. Prenyl acetate is saponified using Sodium hydroxide or Sodium carbonate in the presence of aqueous methanol to yield prenyl alcohol (**Figure 9**) [26].

Prenyl alcohol is known as prenol, 3-Methyl-2-butenyl alcohol, 3, 3-Dimethylallyl alcohol, 3-Methyl-2-buten-1-ol, 3-Methyl-2-butenol, butenol methyl, Methyl-3-but-2-en-1-ol. Prenol is used in the manufacturing industries as an intermediate in pharmaceutical companies and aroma compounds it is also used in the manufacture of Vitamins A and E, the anti-acne drugs tretinoin and isotretinoin [27–29].

Formation of citral is the application of prenol. Citral is an acyclic monoterpene aldehyde with two isoprene units, making it a monoterpene. Citral is a generic word for two geometric isomers, each of which has its own name; the E-isomer is known as Geranial (trans-Citral) or Citral A, or Neral (cis-Citral) is the name given to the

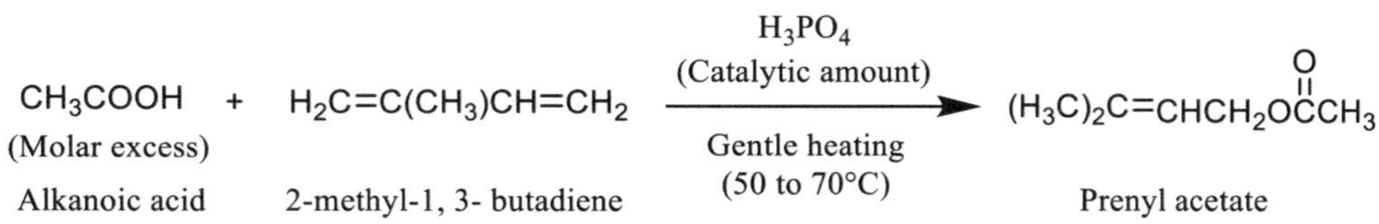

Figure 8.
Preparation of prenyl acetate.

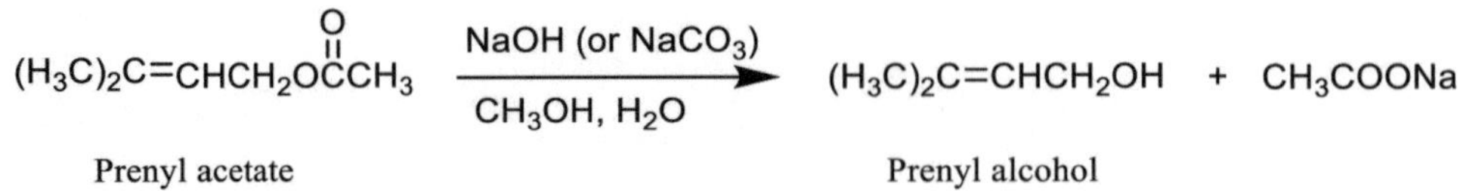

Figure 9.
Formation of prenyl alcohol.

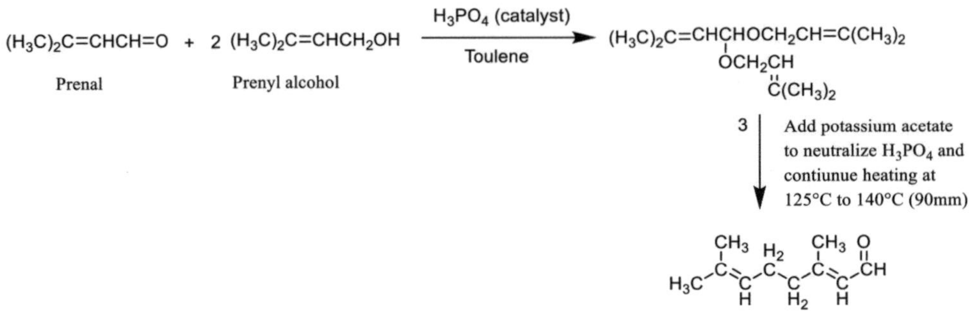

Figure 10.
Synthesis of Citral from prenol by one pot process.

Z-isomer [8, 9]. Citral is used for to hide the smell of smoking; Citral is utilized in the manufacture of vitamin A, lycopene, ionone, and methylionone [30].

Synthesis of Citral from prenol by one pot process: One pot process means it is a method of increasing the effectiveness of a chemical reaction by exposing a reactant to many chemical reactions in a single reactor. Chemists prefer this because it saves time and money while enhancing chemical yield by eliminating a long separation procedure and purification of intermediate chemical compounds (**Figure 10**) [31].

Formation of prenylated indole derivatives: Prenylated indole alkaloids are a vast class of fungal secondary metabolites that include cytotoxic, insecticidal, and antifungal properties. Antibacterial and anthelmintic properties make them a growing subject of study for synthetic and biological researchers [32].

6.3 Prenyl sulfate

Geranyl sulfate, triethylammonium salt and Neryl sulfate, triethylammonium salt these two-reagent system was created for the S-Prenylation of thiols. As alkylating reagents, prenyl sulfate was utilized, allowing the S-Prenylation processes to be carried out in aqueous solutions at room temperature [33].

Synthesis of Geranyl sulfate and Neryl sulfate: Geraniol and Nerol (C10-Prenol) treated with the molar excess of pyridine-SO_3 complex with triethylamine the intended prenyl sulfate precipitated together with the non-consumed prenols when the reaction mixtures were treated with water. The latter were subsequently separated from the residue using cold n-heptane to get yield of triethylammonium salts of pure geranyl sulfate (75 %) and neryl sulfate (72 %) (**Figure 11**) [34].

6.4 Application of Geranyl sulfate and neryl sulfate

Synthesis of benzenethiol derivative: Benzene thiol is known as thiophenol. The thiophenolate is very nucleophilic, implying a high rate of alkylation [35]. Thiophenol

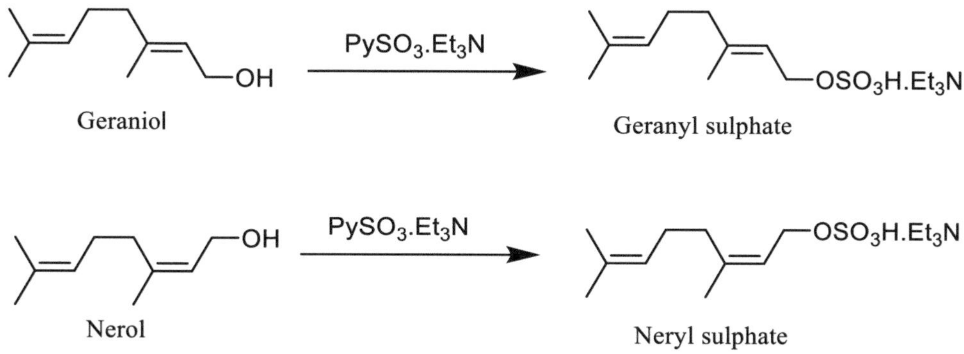

Figure 11.
Synthesis of prenylated sulfates.

are used to make medications such as sulfonamides. Butoconazole and Merthiolate, both antifungal medicines, are thiophenol derivatives [36].

Synthesis of Geranyl cysteine: Reaction proceeds through S-Prenylation. When L-cysteine is treated with sulfate in the presence of aq. KOH then reaction mixture was extracted with diethyl ether to remove triethylamine and geraniol (which was generated, most likely, due to incomplete hydrolysis of sulfate), as per usual S-prenylation technique. It was simple to purify and isolate the required geranyl cysteine as a potassium salt 43% yield (**Figure 12**) [37].

6.5 Prenyl benzaldehyde

The IUPAC name of prenyl benzaldehyde is 4-Chloro-5-(1, 1-dimethylallyl)-2-methoxybenzaldehyde. The synthesized benzaldehydes are being utilized to make a variety of new analogues of Licochalcone A, a well-known antibacterial molecule, as well as to investigate the pharmacophoric components required for antibacterial action (**Figure 13**) [38].

Application of prenyl benzaldehyde: The prenylated benzaldehydes were employed to discover the pharmacophoric components responsible for Licochalcone A as an antibacterial action. It was discovered that the hydroxyl group at position 4′ of the A ring was important for action [39].

6.6 Prenyl bromide

Prenyl bromide is also known as 3, 3-Dimethylallyl bromide, 1-Bromo-3-methyl-2-butene, and Prenyl bromide (**Figure 14**).

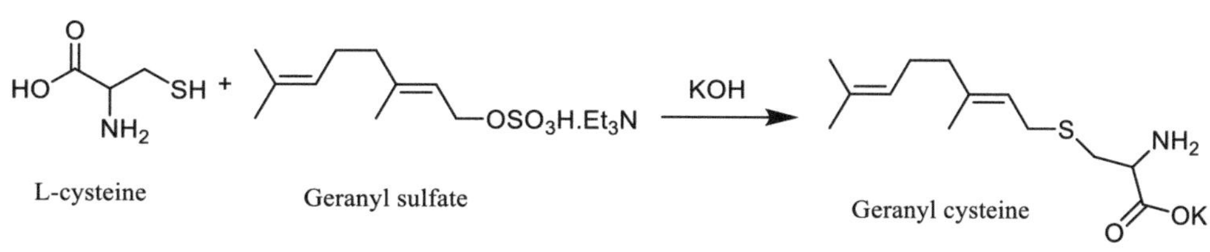

Figure 12.
Synthesis of Geranyl cysteine.

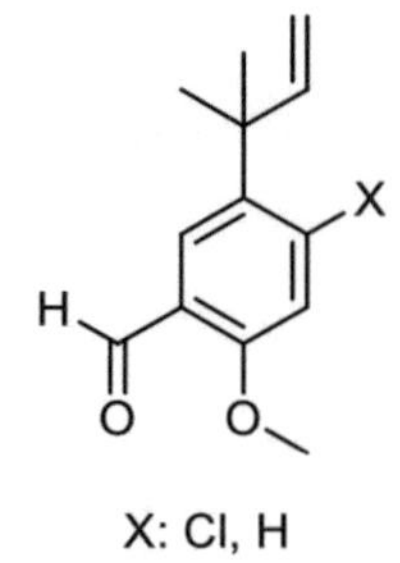

Figure 13.
Structure of prenyl Benzaldehyde.

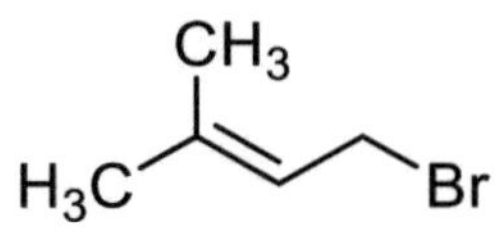

Figure 14.
Structure of Prenyl Benzaldehyde.

Prenyl bromide is soluble water and other organic solvents like ether and acetone. Propyl bromide (n-PB) is used as an intermediate in organic synthesis and in the manufacture of Agrochemicals and Pharmaceuticals. N-PB and n-PB formulations are also used as organic cleaning solvent for precision cleaning, degreasing, electronics and metal cleaning applications [40].

Applications:

Synthesis of N-Prenylation of (-)-indolactam-V: N-prenylation with an electrophilic group. Deprotonation of N-unsubstituted indoles with NaH or, more rarely, KH in DMF, DMSO, THF, or acetone, followed by reaction with prenyl bromide or chloride, is a common method for making N-prenyl indole derivatives. Yields are typically greater than 80%. If electron withdrawing groups (Ac, CHO) were present at C3, alkali hydroxides in the presence of crown ethers, phase transfer catalysts, or K_2CO_3 were utilized as bases. When dealing with NaH/DMF, N-unsubstituted diketopiperazines and phthalimide protected tryptophan derivatives underwent partial epimerization. Alternatively, the disodium salt of tryptophan could be used to selectively N-prenylate unprotected tryptophan by treating it with prenyl bromide/ Na in liquid ammonia (**Figure 15**) [41].

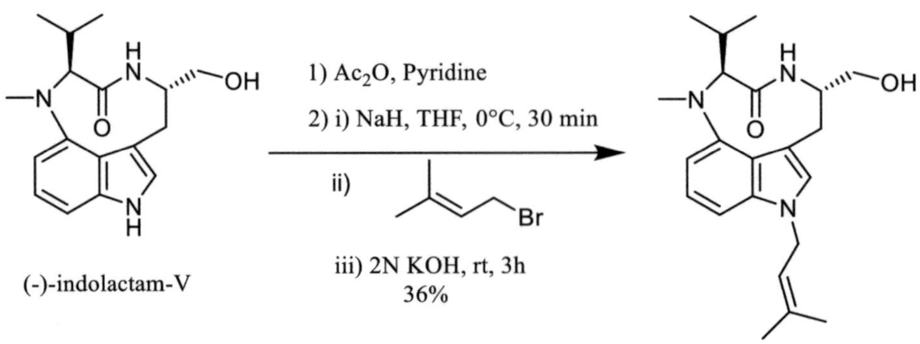

Figure 15.
N-Prenylation of (-)-indolactam.

A series of 5-arylidene-2, 4-thiazolidinediones and its geranyloxy or prenyloxy derivative were synthesized by using Prenyl bromide.

Flavonoids synthesize with prenyl bromide chains, in the presence of anhydrous potassium carbonate using two different methodologies, one based on conventional thermal heating and other by MAOS.

Total Synthesis of (–) Rosiridol and (–)-Bifurcadiol by Catalytic Asymmetric Aldehyde Prenylation:

Synthesis of (-) Rosiridol: Primary alcohol is treated with trityl group as protecting group and triethlamine in the presence of DMAP (acyl transfer reagent) yield compound. The compound undergoes standard allylic oxidation to form alcohol. The obtained compound treated with MnO_2 gives substrate. The catalytic asymmetric prenylation of compound treated with prenyl boronic ester was undergoes the optimized conditions to form homoprenyl alcohol. Removing the protecting group of reaction gives the (-) Rosiridol (**Figure 16**) [42, 43].

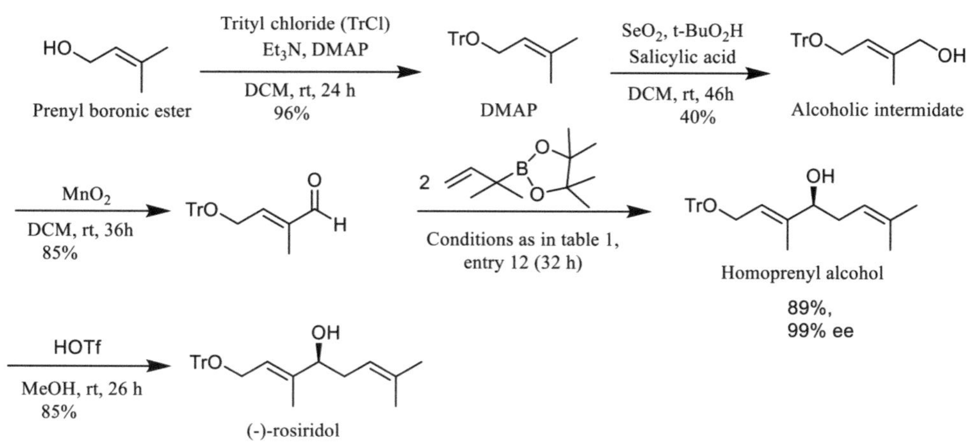

Figure 16.
Synthesis of (-) rosiridol.

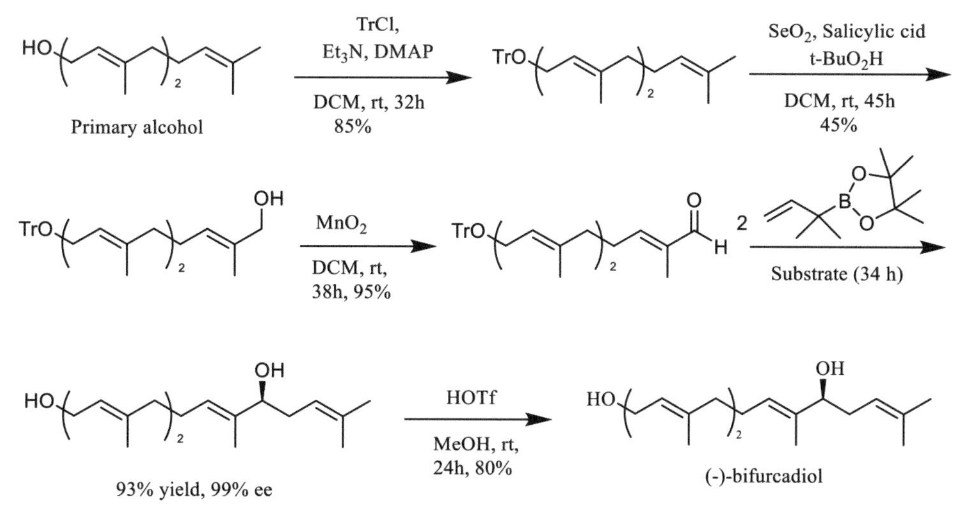

Figure 17.
Synthetic route of (–) Bifurcadiol.

Synthetic route of (–) Bifurcadiol: Primary alcohol is treated with trityl group as protecting group and triethlamine in the presence of DMAP (acyl transfer reagent) yield compound. The compound undergoes standard allylic oxidation to form alcohol. The obtained compound treated with MnO_2 gives substrate. The catalytic asymmetric prenylation of compound treated with prenyl boronic ester was undergoes the optimized conditions to form homoprenyl alcohol. Removing the protecting group of reaction gives the (-) Bifurcadiol (**Figure 17**) [44].

Biogenetic type synthesis of chromanochalcones from prenylated chalcones: By regioselective cyclization using BF_3–Et_2O, we have established a simple and convenient process for the synthesis of chromanochalcones from prenylated chalcones in high yields. Use of BF_3–Et_2O to form a complex with the chelated hydroxyl group (C-2) and α, b-unsaturated carbonyl group of the prenylated chalcone for regioselective cyclization. The production of a flavanone is prevented by this complexation. The chromanochalcone is formed by cyclization of the prenyl group by the second mole of the reagent (**Figure 18**) [45].

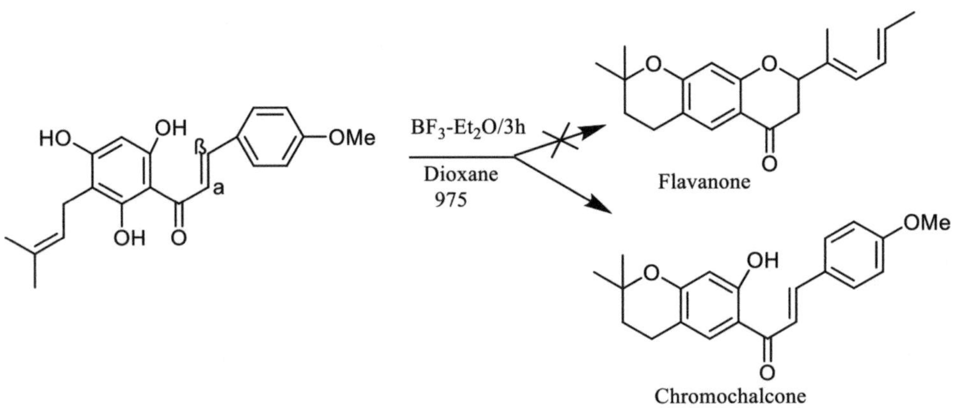

Figure 18.
Biogenetic type synthesis of chromanochalcones from prenylated chalcones using BF_3-Et_2O.

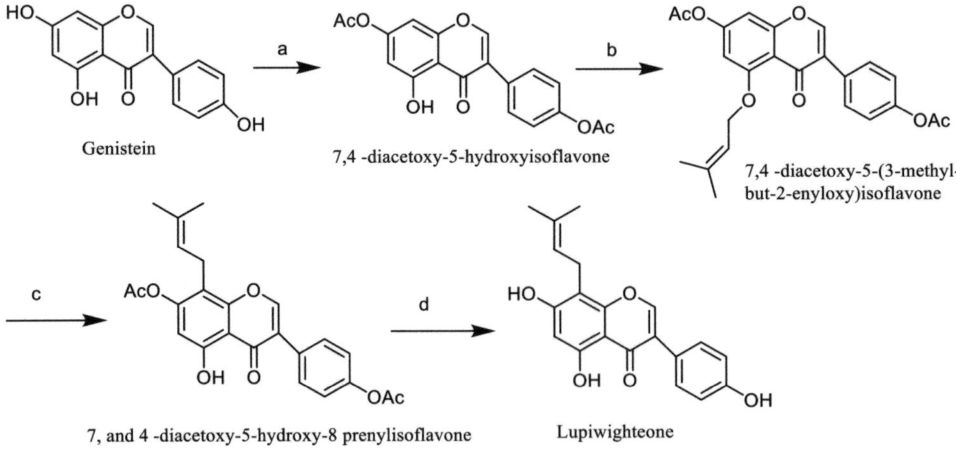

Figure 19.
Synthesis of lupiwighteone.

Synthesis of lupiwighteone: The presence of a prenylated side chain (i.e. prenyl, geranyl) on the flavonoid skeleton separates prenylated flavonoids from other naturally occurring flavonoids. Claisen rearrangements are a highly effective approach for regioselective prenylation of isoflavonones compounds.

The 5-hydroxy group in genistein was treated with the two equivalents of acetic anhydride in pyridine gives 7, 4 -diacetoxy-5-hydroxyisoflavone, when condensation of 7, 4 -diacetoxy-5-hydroxyisoflavone with 3-methyl-2-buten-1-ol in the presence of triphenylphosphine and diethyl azodicarboxylate in dry THF will gives the desired ether 7, 4 -diacetoxy-5-(3-methyl-but-2-enyloxy)isoflavone. The compound 7, 4 -diacetoxy-5-(3-methyl-but-2-enyloxy) isoflavone was dissolved in minimum dry $CHCl_3$ solvent and heated at 60°C. Will form the para-rearrangement product, 7, and 4 -diacetoxy-5-hydroxy-8 prenyl isoflavone, this compound get hydrolysis with 10% aqueous $NaHCO_3$ at 60°C gives the lupiwighteone (**Figure 19**) [46].

7. Biotechnology application

The ability of FTase to change a single cysteine residue in the C-terminal CaaX motif and incorporate isoprenoid analogs with bioorthogonal functions has been utilized for site-specific protein modifications in recent years. This is feasible because the presence of a CaaX-box at the C-terminus of practically any protein is enough to make it an effective FTase substrate. Functionalization of the resultant proteins by bioorthogonal processes provides a simple way for producing a large range of site-specific protein conjugates.

For immobilization of proteins (GFP or GST) onto solid surfaces such as glass slides or agarose resin, both the Poulter and Distefano groups employed azide- and alkyne-functionalized FPP analogs in FTase-catalyzed reactions followed by click reactions or Staudinger ligations. Maynard and colleagues used a similar technique to immobilize mCherry protein tagged with 25 onto a patterned azide-functionalized surface produced by microcontact printing. Waldmann and colleagues used a photochemical thiol-ene reaction between farnesylated recombinant proteins and surface-exposed thiols from functionalized surfaces to immobilize functional proteins in an orientated and selective manner (mCherry and Ypt1) Poulter and colleagues have immobilized the glutathione S-transferase enzyme and antibody-binding protein G to self-assembled monolayers on gold surfaces in a highly organized, regioselective manner. They also developed immobilized recombinant antibody-binding protein L sandwich antibody arrays for trapping antibodies for direct and sandwich-type immunofluorescent detection of ligands in a microarray manner. In general, where oriented protein immobilization is required, the prenylation-based immobilization strategy has several potential biomedical and biotechnology applications, such as protein arrays and diagnostic applications based on immunoassays, Surface Plasmon Resonance (SPR), or electrochemical methods.

Alexandrov and colleagues used a fluorescent analog of FPP and phase partitioning to establish a simple and effective approach for the derivatization and purification of recombinant proteins (such as YPT7, Rab7, and GST). Distefano and colleagues recently reported using an aldehyde functionalized FPP analog in combination with a hydrazide resin-based catch-and-release technique to purify and functionalize proteins containing groups such as a fluorophore or PEG moiety.

The production of site-specific protein modifications such as protein-DNA conjugates, PEGylated proteins, and dually labeled proteins is one key use of the prenylation-based labeling method. A nanoscale-sized defined tetrahedron architecture composed of four oligonucleotides and four GFP molecules, therapeutically relevant proteins GIP and HIV NC attached to oligonucleotides, and DNA-protein cross-links as DNA lesions to study DNA repair and replication are just a few of the protein-DNA conjugates that have been synthesized using this method.

Prenylation of a recombinant protein called ciliary neurotrophic factor (CNTF) with an isoprenoid analog modified with an aldehyde group, followed by oxime ligation-based catch-and-release, resulted in PEGylated CNTF, with the PEG group potentially increasing the serum half-life of this biomedically important protein. Rashidian et al. describe multifunctional macromolecular protein self-assembly made up of an antibody nanoring structure with a single chain anti-CD3 antibody as the targeting element, as well as a model cargo protein and a fluorophore, in a separate paper. The essential multifunctional fragment comprising of a cargo protein, fluorophore, and protein dimerizer was created using a trifunctional FPP analog including both aldehyde and alkyne activity. This high-avidity "effector-antibody-fluorophore" combination was endocytosed by T-leukemia cells, indicating that it might be useful in the development of protein-drug conjugates for therapeutic protein administration and monitoring [47].

8. Conclusions

In conclusion, we have described the prenylated reagent as important for enhancing the protein binding and membrane permeability of compounds. Prenyl alcohol, prenyl ester, prenyl acetate, prenyl bromide, prenyl benzaldehyde and prenyl sulfate are act as a prenyl side chain in the reaction. Prenyl bromide is applicable in synthesis (-) indolactum for their pharmacological studies. (−)-Rosiridol inhibits monoamine oxidase B (MAO B), which is included in neurodegenerative diseases and (−)bifurcadiol displays anti-ulcer and anti-tumor activities. Prenyl esters are utilized to make prenyl alcohol both are used in synthesis of citral by one-pot process and prenylated indole derivatives. Prenylated chalcone is used for the synthesis of chromochalcone because it is an easy and appropriate method for high yield of the compound and catalyzed para-Claisen-Cope rearrangement represents an excellent method for the synthesis of lupiwighteone with the help of a prenylated side chain and gives a simple and high yielding procedure. In the further studies, lot of prenylated molecules developed with the help of a prenylated reagent for giving a better therapeutic effect.

Acknowledgements

The authors thank the Principal and Department of Pharmaceutical Chemistry Mahatma Gandhi Vidyamandir Pharmacy College, Panchvati, Nashik. The authors would also like to thank the Principal and Secretary of Shreeshakti Shaikshanik Sanstha's Divine College of Pharmacy, Satana, Nashik.

Authors' contributions

All the authors have equally contributed to the article. All authors read and approved the final manuscript.

Conflict of interest

The authors have no conflicts of interest.

Acronyms and abbreviations

DMF	Dimethyl formamide
DMSO	Dimethyl sulfoxide
DMAP	4-Dimethylaminopyridine
FPP	Farnesyl diphosphate
FTase	Farnesyl transferase
GGTase-I	Geranyl geranyl transferase-I
GGTase-II or Rab geranylgeranyl transferase	Geranylgeranyl transferase type 2
GGPP	Geranylgeranyl diphosphate
MAOS	Microwave assisted organic synthesis
KH	Potassium hydride
NaH	Sodium hydride
RAM	Ribosomal ambiguity
KOH	Potassium hydroxide
THF	Tetrahydrofuran

Author details

Khemchand R. Surana[1*], Ritesh B. Pawar[2], Ritesh A. Khairnar[3] and Sunil K. Mahajan[4]

1 Department of Pharmaceutical Chemistry, Shreeshakti Shaikshanik Sanstha, Divine College of Pharmacy, Satana, Nashik, Maharashtra, India

2 Department of Pharmaceutical Chemistry, Poona College of Pharmacy, Pune, Maharashtra, India

3 St. John Institute of Pharmacy and Research, Palghar, Maharashtra, India

4 Department of Pharmaceutical Chemistry, MGV's Pharmacy College, Panchvati, Nashik, India

*Address all correspondence to: khemchandsurana411@gmail.com

References

[1] Zhang FL, Casey PJ. Protein prenylation: Molecular mechanisms and functional consequences. Annual Review of Biochemistry. 1996;**65**(1): 241-269. DOI: 10.1146/annurev.bi.65.070196.001325

[2] Amaya M, Baranova A, van Hoek ML. Protein prenylation: A new mode of host–pathogen interaction. Biochemical and Biophysical Research Communications. 2011;**416**(1-2):1-6. DOI: 10.1016/j.bbrc.2011.10.142

[3] Palsuledesai CC, Distefano MD. Protein prenylation: Enzymes, therapeutics, and biotechnology applications. ACS Chemical Biology. 2015;**10**(1):51-62. DOI: 10.1021/cb500791f

[4] Valliere MA, Korman TP, Woodall NB, Khitrov GA, Taylor RE, Baker D, et al. A cell-free platform for the prenylation of natural products and application to cannabinoid production. Nature Communications. 2019;**10**(1):1-9. DOI: 10.1038/s41467-019-08448-y

[5] Alhassan AM, Abdullahi MI, Uba A, Umar A. Prenylation of aromatic secondary metabolites: A new frontier for development of novel drugs. Tropical Journal of Pharmaceutical Research. 2014;**13**(2):307-314. DOI: 10.4314/tjpr.v13i2.22

[6] Kurokawa H, Ambo T, Takahashi S, Koyama T. Crystal structure of Thermobifida fusca cis-prenyltransferase reveals the dynamic nature of its RXG motif-mediated inter-subunit interactions critical for its catalytic activity. Biochemical and Biophysical Research Communications. 2020;**532**(3):459-465. DOI: 10.1016/j.bbrc.2020.08.062

[7] Rodríguez-Concepción M, Boronat A. Isoprenoid biosynthesis in prokaryotic organisms. In: Isoprenoid Synthesis in Plants and Microorganisms. New York, NY: Springer; 2012. pp. 1-16. DOI: 10.1007/978-1-4614-4063-5_1

[8] Gosser YQ, Nomanbhoy TK, Aghazadeh B, Manor D, Combs C, Cerione RA, et al. C-terminal binding domain of Rho GDP-dissociation inhibitor directs N-terminal inhibitory peptide to GTPases. Nature. 1997;**387**(6635):814-819. DOI: 10.1038/42961

[9] Casey PJ. Biochemistry of protein prenylation. Journal of Lipid Research. 1992;**33**(12):1731-1740. DOI: 10.1016/S0022-2275(20)41331-8

[10] Yokoyama K, McGeady P, Gelb MH. Mammalian protein geranylgeranyltransferase-I: substrate specificity, kinetic mechanism, metal requirements, and affinity labeling. Biochemistry. 1995;**34**(4):1344-1354. DOI: 10.1007/978-3-540-49755-4_29

[11] Placzek AT, Krzysiak AJ, Gibbs RA. Chemical probes of protein prenylation. In: The Enzymes. Vol. 30. Academic Press; 2011. pp. 91-127. DOI: 10.1016/B978-0-12-415922-8.00005-7

[12] Patel DV, Schmidt RJ, Biller SA, Gordon EM, Robinson SS, Manne V. Farnesyl diphosphate-based inhibitors of Ras farnesyl protein transferase. Journal of Medicinal Chemistry. 1995;**38**(15):2906-2921. DOI: 0022-2623/95/1838-2906$09

[13] Ashby MN, Edwards PA. Elucidation of the deficiency in two yeast coenzyme Q mutants. Characterization of the structural gene encoding

hexaprenyl pyrophosphate synthetase. Journal of Biological Chemistry. 1990;**265**(22):13157-13164. DOI: 10.1016/S0021-9258(19)38280-8

[14] Kral AM, Diehl RE, deSolms SJ, Williams TM, Kohl NE, Omer CA. Mutational analysis of conserved residues of the-subunit of human farnesyl: Protein transferase. Journal of Biological Chemistry. 1997;**272**(43):27319-27323. DOI: 10.1074/jbc.272.43.27319

[15] Lane KT, Beese LS. Thematic review series: lipid posttranslational modifications. Structural biology of protein farnesyltransferase and geranylgeranyltransferase type I. Journal of Lipid Research. 2006;**47**(4):681-699. DOI: 10.1194/jlr.R600002-JLR200

[16] Huang CC, Casey PJ, Fierke CA. Evidence for a catalytic role of zinc in protein farnesyltransferase: Spectroscopy of Co2+-farnesyltransferase indicates metal coordination of the substrate thiolate. Journal of Biological Chemistry. 1997;**272**(1):20-23. DOI: 10.1074/jbc.272.1.20

[17] Harris CM, Poulter CD. Recent studies of the mechanism of protein prenylation. Natural Product Reports. 2000;**17**(2):137-144. DOI: 10.1039/A904110I

[18] Cassidy PB, Poulter CD. Transition state analogs for protein farnesyltransferase. Journal of the American Chemical Society. 1996;**118**(36):8761-8762. DOI: 10.1021/ja961214c

[19] Weller VA, Distefano MD. Measurement of the alpha-secondary kinetic isotope effect for a prenyltransferase by MALDI mass spectrometry. Journal of the American Chemical Society. 1998;**120**(31):7975-7976. DOI: 10.1021/ja980353m

[20] Mu Y, Omer CA, Gibbs RA. On the stereochemical course of human protein-farnesyl transferase. Journal of the American Chemical Society. 1996;**118**(8):1817-1823. DOI: 10.1021/ja953005i

[21] Reed BC, Rilling HC. Substrate binding of avian liver prenyltransferase. Biochemistry. 1976;**15**(17):3739-3745. DOI: 10.1021/bi00662a015

[22] Tan DH, Zeng YF, Liu Y, Lv WX, Li Q, Wang H. Direct assembly of prenylated heteroarenes through a cascade minisci reaction/dehydration sequence. ChemistryOpen. 2016;**5**(6):535. DOI: 10.1002/open.201600096

[23] Basuli S, Sahu S, Saha S, Maji MS. Cp* Co (III)-Catalyzed dehydrative C2-Prenylation of pyrrole and indole with Allyl alcohols. Advanced Synthesis & Catalysis. 2021;**363**(19):4605-4611. DOI: 10.1002/adsc.202100811

[24] Babler JH, inventor; Loyola University Chicago, assignee. Methods for conversion of isoprene to prenyl alcohol and related compounds. United States patent US 6,278,016. 2001 Aug 21. Patent No.: US 6,278,016 B1

[25] Holscher W, Vitzthum OG, Steinhart H. Prenyl alcohol-source for odorants in roasted coffee. Journal of Agricultural and Food Chemistry. 1992;**40**(4):655-658. DOI: 10.1007/BF01132293

[26] Ruan FX, He ZP, Song HL, Huang KL, Hu J, Peng XY, et al. The catalyzed synthesis of prenyl alcohol in anhydrous system. In: Advanced Materials Research. Vol. 881. Trans Tech Publications Ltd.; 2014. pp. 241-244. DOI: 10.4028/www.scientific.net/AMR.881-883.241

[27] Jiang X, Zhang Y, Li Y, Zhang Q, Yin H. Synthesis of prenyl alcohol

from isoprene. In: 2015 International Symposium on Energy Science and Chemical Engineering. Atlantis Press; 2015. pp. 108-111. DOI: 10.2991/isesce-15.2015.22

[28] Boronat Rigol A, Soldevila-Domenech N, Rodríguez-Morató J, Martínez-Huélamo M, Lamuela-Raventós RM, Torre Fornell RD. Beer phenolic composition of simple phenols, prenylated flavonoids and alkylresorcinols. Molecules. 2020;**25**(11):E2582. DOI: 10.3390/molecules25112582

[29] Dubey NK, Takeya K, Itokawa H. Citral: A cytotoxic principle isolated from the essential oil of Cymbopogon citratus against P388 leukaemia cells. Current Science. 1997;**73**(1):22-24. DOI: https://www.jstor.org/stable/24098141

[30] Han X, Zhou SJ, Tan YZ, Wu X, Gao F, Liao ZJ, et al. Crystal structures of saturn-like C50Cl10 and pineapple-shaped C64Cl4: Geometric implications of double-and triple-pentagon-fused chlorofullerenes. Angewandte Chemie International Edition. 2008;**47**(29):5340-5343. DOI: 10.1002/anie.200800338

[31] Tanaka S, Shiomi S, Ishikawa H. Bioinspired indole prenylation reactions in water. Journal of Natural Products. 2017;**80**(8):2371-2378. DOI: 10.1021/acs.jnatprod.7b00464

[32] Maltsev S, Sizova O, Utkina N, Shibaev V, Chojnacki T, Jankowski W, et al. Prenyl sulfates as alkylating reagents for mercapto amino acids. Acta Biochimica Polonica. 2008;**55**(4):807-813. DOI: 10.18388/abp.2008_3044

[33] Favre HA, Powell WH. Nomenclature of organic chemistry: IUPAC recommendations and preferred names 2013. Royal Society of Chemistry. 2013. DOI: 10.1039/9781849733069-00552

[34] Wakselman M. Di-t-butyl Dicarbonate. Encyclopedia of Reagents for Organic Synthesis. 2001. DOI: 10.1002/047084289X.rd060

[35] Khazaei A, Kazem-Rostami M, Moosavi-Zare AR, Bayat M, Saednia S. Novel one-pot synthesis of thiophenols from related triazenes under mild conditions. Synlett. 2012;**23**(13):1893-1896. DOI: 10.1055/s-0032-1316557

[36] Rosado JA, Sage SO. Regulation of plasma membrane Ca2+-ATPase by small GTPases and phosphoinositides in human platelets. Journal of Biological Chemistry. 2000;**275**(26):19529-19535. DOI: 10.1074/jbc.M001319200

[37] Besouw M, Masereeuw R, van den Heuvel L, Levtchenko E. Cysteamine: An old drug with new potential. Drug Discovery Today. 2013;**18**(15-16):785-792. DOI: 10.1016/j.drudis.2013.02.003

[38] Kromann H, Larsen M, Boesen T, Schønning K, Nielsen SF. Synthesis of prenylated benzaldehydes and their use in the synthesis of analogues of licochalcone A. European Journal of Medicinal Chemistry. 2004;**39**(11):993-1000. DOI: 10.1016/j.ejmech.2004.07.004

[39] Khanna RN, Singh PK. Prenylation of quinonoid compounds with prenyl bromide using lead bromide/aluminum powder as catalyst. Synthetic Communications. 1990;**20**(12):1743-1749. DOI: 10.1080/00397919008053098

[40] Lindel T, Marsch N, Adla SK. Indole prenylation in alkaloid synthesis. Alkaloid Synthesis. 2011:67-129. DOI: 10.1007/128_2011_204

[41] Hossain SU, Bhattacharya S. Synthesis of O-prenylated and O-geranylated derivatives of 5-benzylidene2, 4-thiazolidinediones and evaluation of

their free radical scavenging activity as well as effect on some phase II antioxidant/detoxifying enzymes. Bioorganic & medicinal chemistry letters. 2007;**17**(5):1149-1154. DOI: 10.1016/j.bmcl.2006.12.040

[42] Neves MP, Cidade H, Pinto M, Silva AM, Gales L, Damas AM, et al. Prenylated derivatives of baicalein and 3, 7-dihydroxyflavone: Synthesis and study of their effects on tumor cell lines growth, cell cycle and apoptosis. European Journal of Medicinal Chemistry. 2011;**46**(6):2562-2574. DOI: 10.1016/j.ejmech.2011.03.047

[43] Zhang YL, Zhao ZN, Li WL, Li JJ, Kalita SJ, Schneider U, et al. Catalytic asymmetric aldehyde prenylation and application in the total synthesis of (−)-rosiridol and (−)-bifurcadiol. Chemical Communications. 2020;**56**(69): 10030-10033. DOI: 10.1039/D0CC00367K

[44] Narender T, Reddy KP. BF3–Et2O mediated biogenetic type synthesis of chromanochalcones from prenylated chalcones via a regioselective cyclization reaction. Tetrahedron Letters. 2007;**48**(43):7628-7632. DOI: 10.1016/j.tetlet.2007.08.124

[45] Mula S, Patro BS, Kalena GP, Chattopadhyay S. Novel synthesis of prenylated Phenols and their antioxidant properties. Natural Product Communications. 2006;**1**(2):1934578X0600100209. DOI: 10.1177%2F1934578X0600100209

[46] Al-Maharik N, Botting NP. Synthesis of lupiwighteone via a para-Claisen–Cope rearrangement. Tetrahedron. 2003;**59**(23):4177-4181. DOI: 10.1016/S0040-4020(03)00579-9

[47] Kim KW, Chung HH, Chung CW, Kim IK, Miura M, Wang S, et al. Inactivation of farnesyltransferase and geranylgeranyltransferase I by caspase-3: cleavage of the common α subunit during apoptosis. Oncogene. 2001;**20**(3):358-366. DOI: 10.1038/sj.onc.1204099

Chapter 5

Prenylation of Natural Products: An Overview

Kantharaju Kamanna and Aravind Kamath

Abstract

Natural products with varied functional attributes are available in large abundance in nature. Nature has been an infinite repository of resources leading to drug development, discovery of novel chemicals, pharmacophores, and several invaluable bioactive agents. Natural products play a critical role in modern drug development, especially for antibacterial and antitumor agents. Their varied chemical structure, composition, solubility, and synthetic pathways bestow upon them a high level of diversity. Prenylation is a covalent addition of hydrophobic moieties to proteins or any other chemical compounds. Generally, the hydrophobic moieties are farnesyl or geranylgeranyl isoprenyl groups. Prenylation of flavonoids, alkaloids, terpernoids, etc., leads to gain of varied functionalities to the natural products in addition to the already existing functions. The ever-increasing need for the discovery of new drugs finds a new avenue through the prenylation of natural products. Cell-free synthesis of the prenylated natural products can be seen as a new alternative for the natural synthesis, which warrants time-consuming isolation and purification techniques.

Keywords: prenylation, natural products, antibacterial, antitumor, alkaloids

1. Introduction

Anything that is produced by life including biotic materials such as silk, hair, bio-based materials such as bioplastics, cornstarch, bodily fluids such as blood, milk, and other natural materials that were once found in living organisms such as shell, soil, coal, can all be called as natural products. They are products from various natural sources such as plants, microbes, and animals. The whole of the organism, a part of an organism, an extract from an organism, or pure compounds isolated from the organisms such as alkaloids, coumarins, flavonoids, steroids, lectins, lignans, terpenoids, nonribosomal polypeptides, and polyketides can all be termed as natural products. A limited scope of the definition of natural product can be any molecule synthesized by a living organism. Organic chemistry as we know today has its roots in the study of natural products. The semisynthetic chemistry is an offshoot of organic chemistry wherein the natural products are modified to alter/improve and enhance their activities.

The natural selection and evolutionary processes over millions of years have bestowed the natural products with high structural diversity and unique pharmacological or biological activities. Natural products exhibit structural diversity that is far exceeding the variety that could be synthesized in a laboratory. Classification of natural products is often based on their biological function, biosynthetic pathway, or their source. Primary metabolites and secondary metabolites are the two major classes of natural products. The substances required for an organism to survive are termed as primary metabolites, whereas the substances that are not required for an organism to survive are termed as secondary metabolites. Secondary metabolites confer the organism with advantage in growth and survival within its environment. In practice, the term natural products generally refers to the secondary metabolites and small molecules with molecular weight < 1500 amu.

Natural products have been used for medicinal purposes since ancient times as herbal remedies. Natural products and their structural analogues have a strong impact on human culture and have been used throughout human history as condiments, pigments, and pharmaceuticals. Many of the natural products are potential drug candidates due to the prevailing increased antibiotic resistance. In comparison to the standard combinatorial chemistry, the natural products provide distinct structural diversity and functions. Limited by the lack of cost-effective production methodologies, the study and therapeutic potential of natural products have not been optimally explored. The similarities in the structures and variation in the sources of isolation make it difficult to isolate the natural products. The challenges associated with isolation/production of natural products are circumvented by development of several semisynthetic chemical syntheses.

Due to the safety and efficacy of the natural products, they have been the drugs of choice in improving the human health despite facing a tough competition from compounds derived from computational and combinatorial chemistry. Their importance in drug discovery has been enhanced owing to their largely untapped structural diversity [1]. Natural products containing prenyl side chains represent a rare class in themselves. For several decades now prenylated natural products are recognized as interesting and valuable biologically active phytochemicals [2]. Simple modifications by biological or chemical approaches produce a variety of prenylated aromatic compounds with added structural diversity, altered biological activity, and enhanced therapeutic potential.

A covalent addition of any hydrophobic moiety to protein or any other chemical compound can be termed as prenylation. In case of proteins, generally it is the addition of farnesyl or geranylgeranyl moiety to the cysteine residue via a thioester linkage at the C-terminus. This addition of prenyl moiety bestows novel hydrophobic properties on proteins that leads to the localization of prenylated proteins to the plasma membrane or organellar membranes. It has been shown that well-characterized prenylated proteins are major players in most of the cell signal transduction pathways.

Prenylation of natural products enhances various biological activities as compared with the respective nonprenylated compounds. Due to their versatile and promising pharmacological properties and health benefits on multitarget tissues, the prenylated forms have gained prominence [3, 4]. The increased lipophilicity of prenylated natural products as compared with nonprenylated forms leads to high affinity with cell membranes and enhanced biological activities or significant pharmacological effects [4, 5]. A multitude of biological activities offered by these compounds justifies their enhanced pharmacological investigation. Recent in-depth investigation of prenylated natural compounds with the prenyl substituents playing a key role in the molecular activity has led to discovery of promising anticancer, anti-inflammatory, antioxidant, and neuroprotective compounds. The prenylation of natural compounds is catalyzed by the several enzyme groups of prenyltransferases (PTases), including membrane-embedded UbiA-type, bacterial and fungal ABBA-type, and fungal dimethylallyl tryptophan synthase (DMATS)-type PTases [6–8].

2. Different classes of natural products

Natural products belong to several different classes of molecules. On the basis of their biosynthetic origin, they can be classified as: alkaloids, phenylpropanoids, polyketides, and terpenoids. Prenyl groups appear in a wide variety of these natural products of microbial and plant origin, including amino acids, stilbenes, alkaloids, polyketides, and phenylpropanoids such as flavonoids, creating natural product hybrids with altered or enhanced bioactivities.

2.1 Alkaloids

The term "alkaloid," introduced in 1819 by the German chemist Carl Friedrich Wilhelm Meißner, is derived from Latin root *alkali* (which, in turn, comes from the *Arabic al-qalwi* meaning potassium-carbonate-containing ashes of plants). Heterocyclic nitrogen-containing compounds biosynthesized from amino acids can be termed as alkaloids, though for reasons historical and/or otherwise, there are many exceptions to this rule. Alkaloids represent one of the biggest classes of natural products, and due to the large number and structural diversity, they offer a vast field of investigation. Based on their biosynthetic precursor and heterocyclic ring system, alkaloids are classified into diverse categories (**Figure 1**), namely indole (**1**), purine (**2**), quinolone (**3**), isoquinoline (**4**), tropane (**5**), imidazole (**6**), etc. [9].

Prenylated indole alkaloids are a large family of secondary metabolites containing indole/indoline and isoprenoid moieties or structures derived thereof. These alkaloids generally contain a diketopiperazine (**7**) or a bicyclo [2.2.2] diazaoctane ring (**8**) as a core structure and are biogenetically derived from tryptophan (**9**) (**Figure 2**), a cyclic amino acid, and one or two isoprene units [10]. From filamentous fungi, especially from the genera Penicillium and Aspergillus, numerous prenylated indole alkaloids including asperparalines (**10**), brevianamides (**11**), marcfortines (**12**), notoamides (**13**), paraherquamides (**14**), stephacidins (**15**), and versicolamides (**16**) (**Figure 3**) have been isolated [11]. They are attractive targets for chemical synthesis, biosynthesis, and biological activity studies due to the fact that they exhibit a diverse range of relevant biological activities such as insecticidal, cytotoxic, anthelmintic, and antibacterial properties.

Asterriquinones (**17a** and **17b**) are a large group of the prenylated indole alkaloids containing two tryptophan moieties with a bis (indolyl) benzoquinone structure (**Figure 4**). They exhibit remarkable pharmacological activities such as antiretroviral,

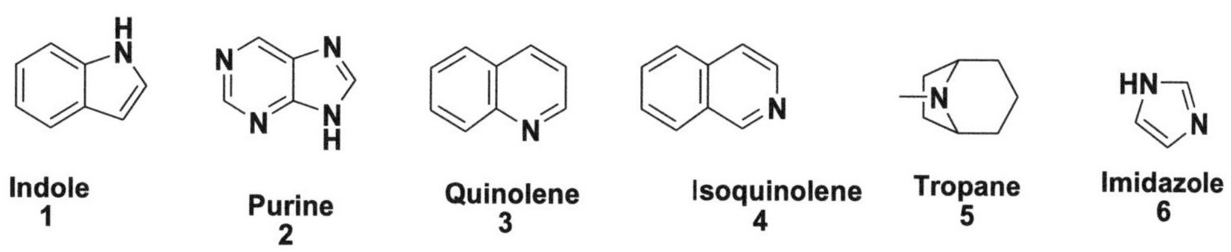

Figure 1.
Diverse categories of alkaloids.

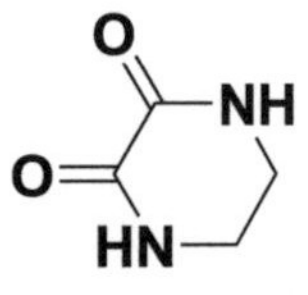

diketopiperazine
7

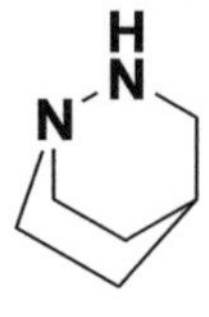

bicyclo [2.2.2] diazaoctane
8

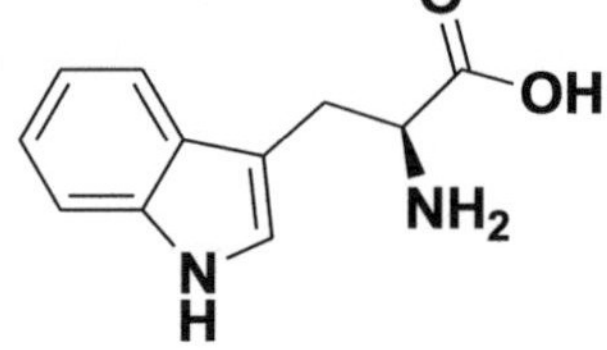

tryptophan
9

Figure 2.
Prenylated indole alkaloids core structure.

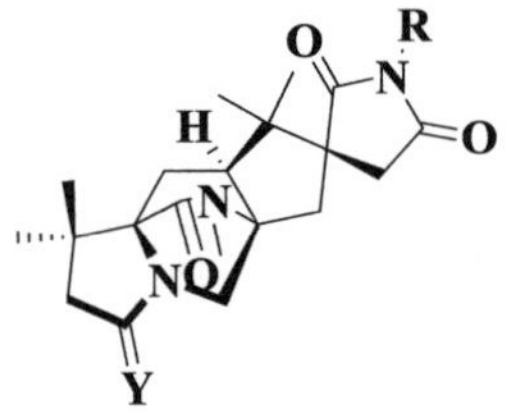

Asperparaline (R=Me,Y=H,H)
10

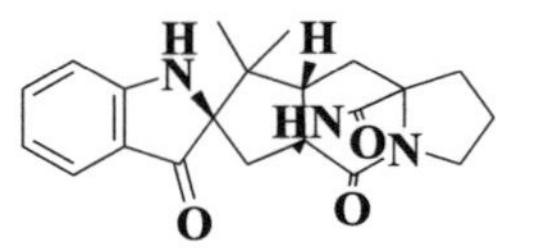

Brevianamide
11

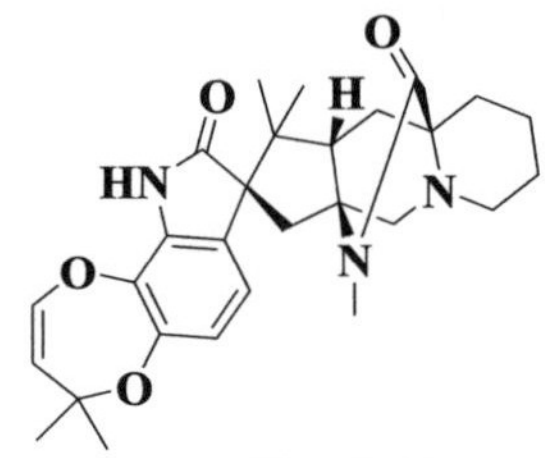

Marcfortine
12

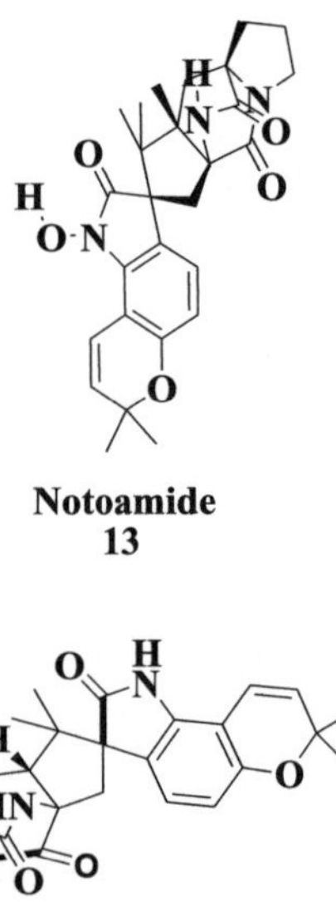

Notoamide
13

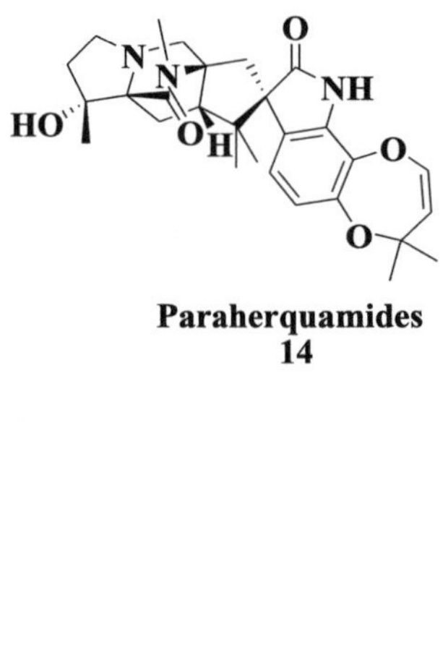

Paraherquamides
14

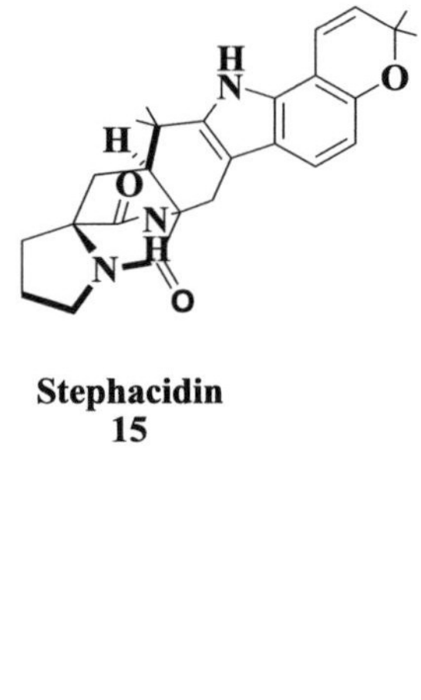

Stephacidin
15

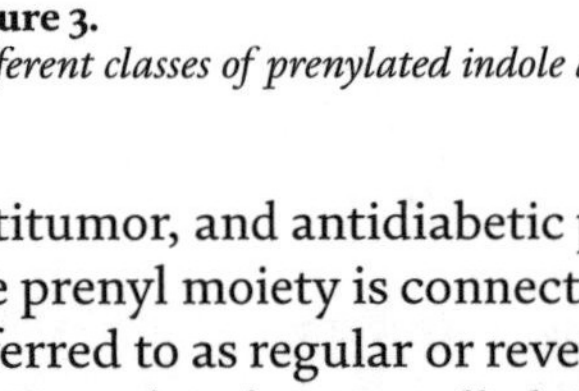

Versicolomide
16

Figure 3.
Different classes of prenylated indole alkaloids.

antitumor, and antidiabetic properties [12]. In prenylated indole alkaloids **(18)**, the prenyl moiety is connected at either C1 or C3 to an aromatic nucleus, which are referred to as regular or reverse prenylation, respectively (**Figure 5**).

Prenylated purine alkaloids isolated from the seeds of *Gleditsia japonica* have been identified as prenylated purine alkaloid glucosides and named as the locustoside **(19)** (**Figure 6**). The plant cytokinin N^6-isopentenyladenine **(20)** (**Figure 7**) and

DOI: http://dx.doi.org/10.5772/intechopen.104636

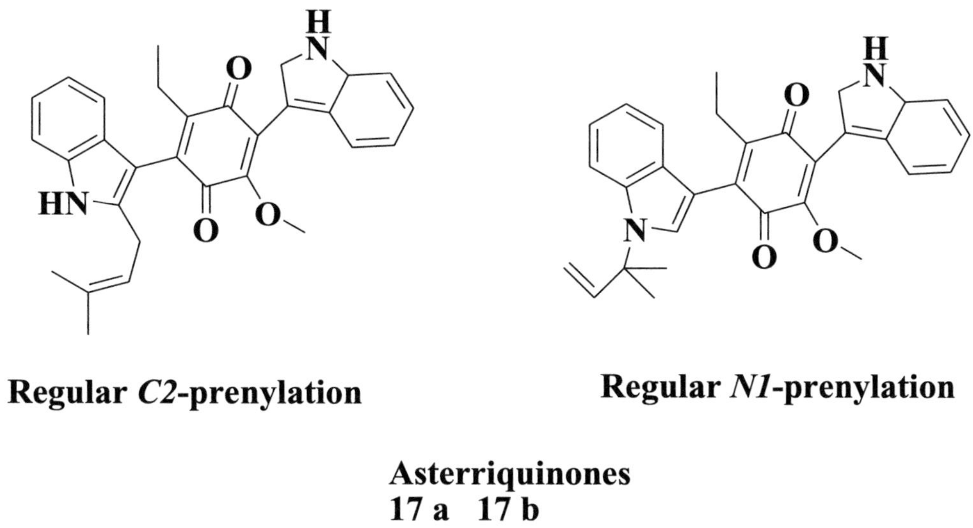

Figure 4.
Asterriquinones with a bis (indolyl) benzoquinone structure.

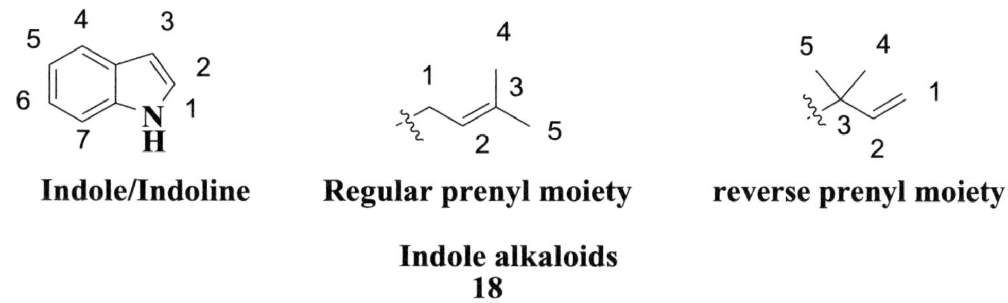

Figure 5.
Regular or reverse prenylation of indole alkaloids.

N^3-prenylated purine alkaloids, e.g., triacanthine **(21)** (**Figure 7**), from the leaves of *Gleditsia triacanthos* have been reported [13, 14]. Triacanthine **(21)** shows hypertensive activity, also cardiotonic, antispasmodic, and a respiratory analeptic. It has also been reported to exert antitumor effects in bladder cancer in vitro and in vivo [15]. The cytokinin, N^6-(Δ^2-isopentenyl) adenine, is found to be 3.3 times as active as N^6-(Δ^2-isopentenyl) adenosine in promoting the growth of cytokinin-requiring tobacco (*Nicotiana tabacum*) callus [16].

Prenylated quinolinone alkaloids, aspoquinolones **(22, 23)** (**Figure 8**), and prenylated isoindolinone alkaloids, aspernidines **(24, 25, 26)** (**Figure 8**), have been isolated and characterized from the fungus *Aspergillus nidulans*. These compounds exhibit varied cytotoxicity against various human cancerous cells. Aspoquinolones differ at the configuration of cyclopropyl ring pointing to the fact that the specific configuration of the cyclopropyl ring is essential for their cytotoxic activity [17].

Prenylated alkaloids isolated from plants and fungi are a good example of high structural diversity from only a limited array of structurally nondiverse starting materials. The assembly of complex carbon skeletons is mediated by enzyme catalyzed selective C–H oxidation reactions. The ambivalent reactivity of the

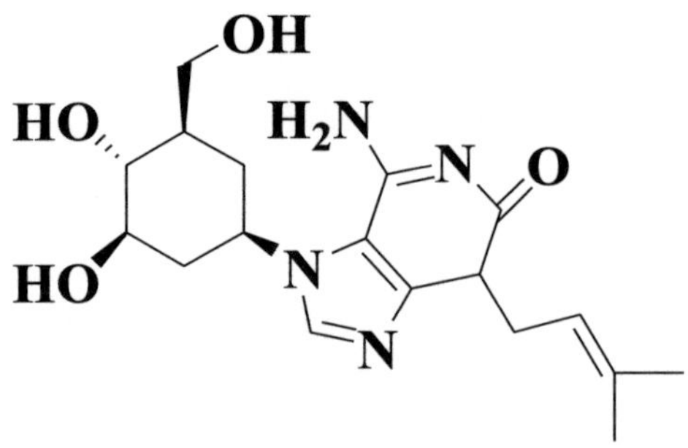

Locustoside
19

Figure 6.
Prenylated purine alkaloid.

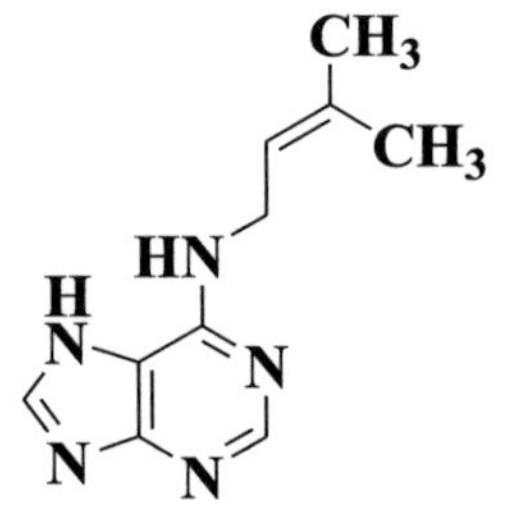

N^6-Isopentenyladenine
20

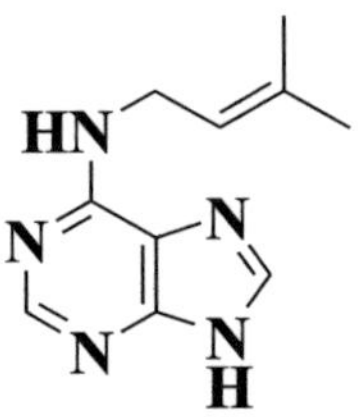

Triacanthine
21

Figure 7.
N^6 and N^3 -prenylated purine alkaloids.

heteroatom is exploited in the diverse condensation chemistry during the prenylated alkaloid biogenesis [18].

2.2 Phenylpropanoids

A diverse family of secondary metabolites synthesized by plants, bacteria, and fungi from the amino acids phenylalanine and tyrosine are termed as phenylpropanoids. The term "phenylpropanoid" is generally used to refer to any compound bearing a 3-carbon propene chain attached to 6-carbon aromatic phenyl ring (C6-C3 compounds). Most of the phenylpropanoids are formed from cinnamic or *p*-coumaric acids. Several pharmacological activities including antimicrobial, antioxidant, anti-inflammatory, antidiabetic, and anticancer activities have been attributed to these diverse groups of compounds that can be found to be present in spices, herbs, fruits, vegetables, and cereal grains. Owing to their antioxidant property, they exhibit renoprotective, neuroprotective, cardioprotective, and hepatoprotective effects [18].

The prenylations of umbelliferone **(27)** in the 6 or 8 position yield demethylsuberosin **(28)** and osthenol **(29)** (**Figure 9**) and give access to the branch pathways to linear or angular furano and pyranocoumarins, which are predominantly found in the *Umbelliferae* [19]. The hydrolysis of the secondary signal messengers' cyclic adenosine monophosphate (cAMP) and cyclic guanosine monophosphate (cGMP) is catalyzed

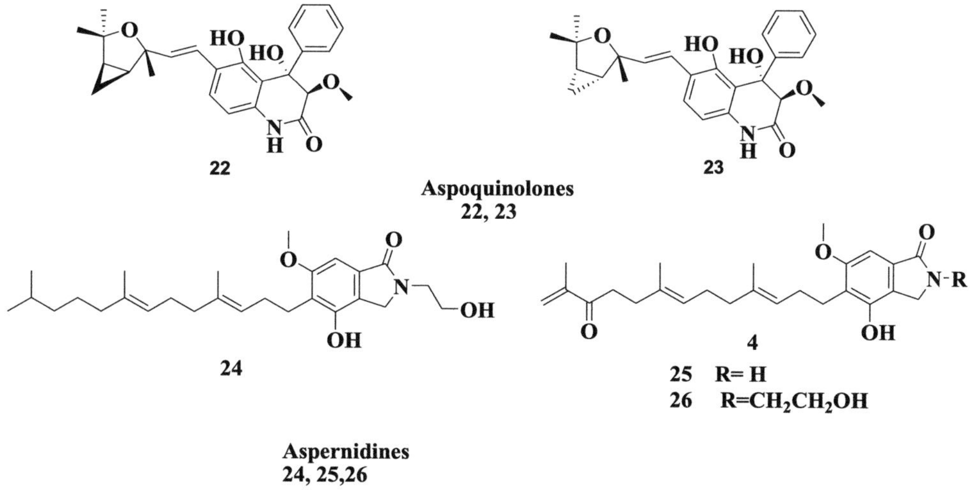

Figure 8.
Prenylated quinolinone alkaloids (22, 23) *and isoindolinone alkaloids* (24, 25, 26).

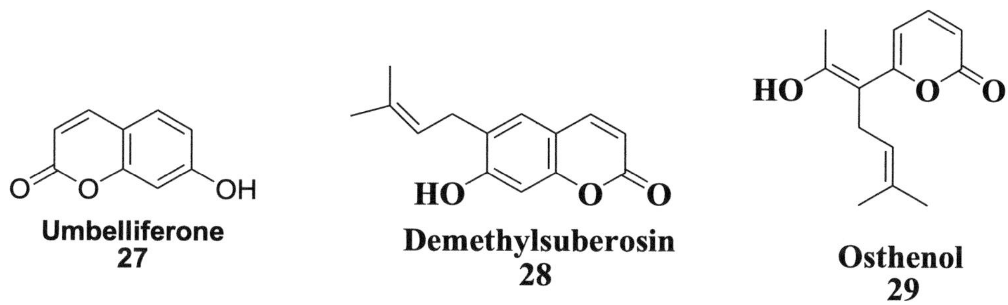

Figure 9.
Prenylated umbelliferone.

by the phosphodiesterases (PDEs). PDE inhibitors are therapeutic targets of high interest for central nervous system (CNS), inflammatory, and respiratory diseases. The prenylated coumarins **(30, 31)** from *Toddalia asiatica* (**Figure 10**) exhibit wide range of inhibition against the PDEs [20].

Marianins are the prenylated phenylpropanoids, isolated from the fungus *Mariannaea camptospora*. Marianin A **(32)** is a 5-methylcoumarin bearing two prenyloxy groups, whereas Marianin B **(33)** is an orcinol derivative substituted with a 3, 3-dimethyl-4-pentenoyl chain (**Figure 11**). Marianins show a weak antimicrobial activity and lack any significant anticancerous activity [21].

Flavonoids are valuable natural phenylpropanoids products and widely distributed in the plant kingdom bestowing a self-defensive strategy to the plants. Flavonoids are categorized on the basis of their oxidative states and substituents, into chalcones **(34)**, flavones **(35)**, flavanones **(36)**, isoflavones **(37)**, dihydroflavonols **(38)**, anthocyanidins **(39)**, etc. [22]. Flavonoids consist of C6-C3-C6 skeleton with two aromatic rings A and B and a (dihydro) benzopyran ring C adjacent to A (**Figure 12**).

Prenylation at the two benzene rings, or α, β carbons in chalcones while enhancing the structural diversity, increases their bioactivities as well. The cytotoxic activity of the 3-hydroxylated derivative of xanthohumol **(40)** is much higher than its

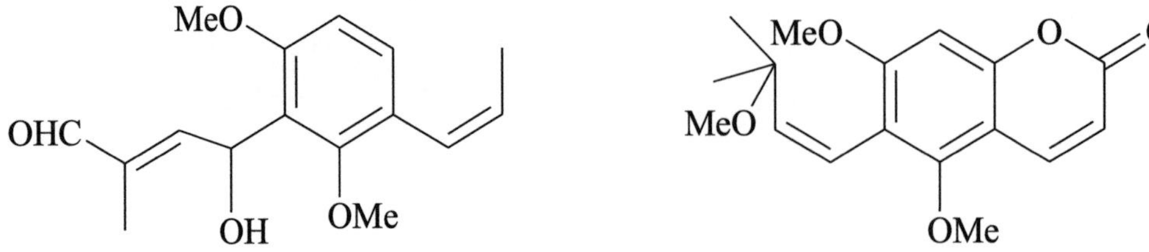

Prenylated coumarins
30, 31

Figure 10.
Prenylated coumarins.

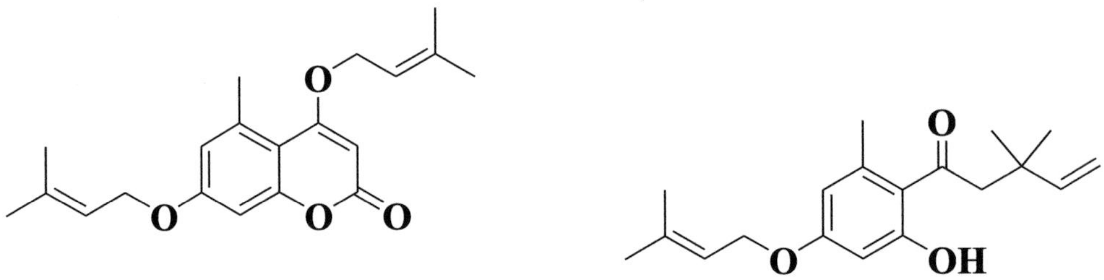

Marianin A
32

Marianin B
33

Figure 11.
Prenylated phenylpropanoid.

nonprenylated analogue 3-hydroxyhelichrysetin **(41)** (**Figure 13**). Diverse biological and pharmacological activities such as antimicrobial and antiviral (C5-prenylated derivatives), antioxidant (the C–H bonds of the prenyl substituents are the most thermodynamically preferred sites for free radical attack, and thus play an important role in the antioxidant activity) cytotoxic, chemopreventive, and estrogenic activities are attributed for prenylated chalcones [23].

The success of cancer therapy is largely impeded by the development of multi-drug resistance (MDR) by tumor cells. The MDR conferred to the cancer cells by the overexpression of the P-glycoprotein (Pgp) [24]. In comparison to the flavanones, isoflavones, and glycosyl derivatives, chalcones, flavones, and flavonols bind more strongly to Pgp cytosolic site. For the ability of these modulators to mimic the adenine moiety of ATP, the hydroxylation at position 5 is essential, in addition to the presence of a ketone at position 4 [25]. Interestingly, the modulating effects of C-prenylated derivatives produced by nontoxic concentrations suggest that these compounds should be investigated in vivo as potential Pgp modulator in tumor cells.

2.3 Polyketides

Polyketides are produced by bacteria, fungi, plants, and few marine organisms. These secondary metabolites exhibit a high degree of structural diversity, even though they are synthesized from simple acyl building blocks. They form a chain of either

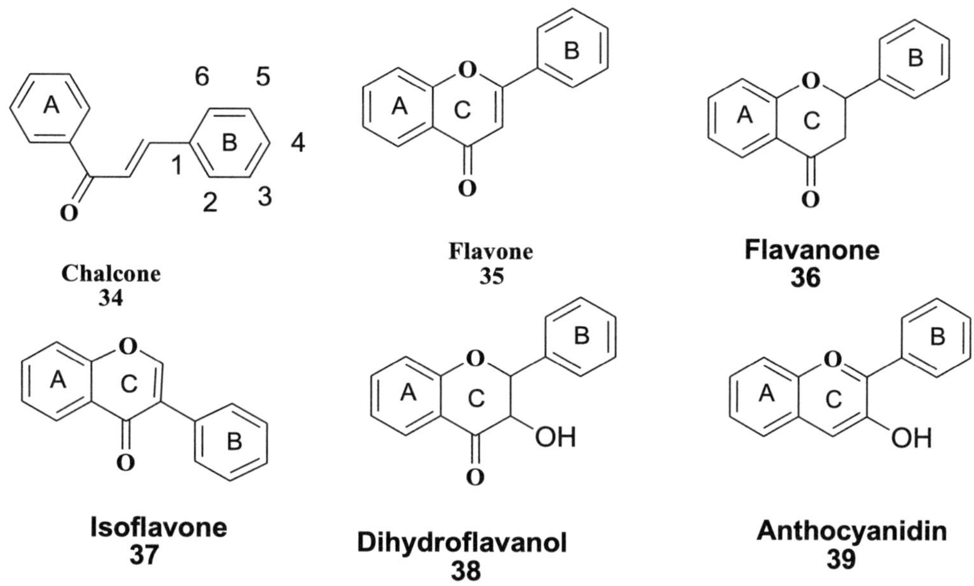

Figure 12.
Flavonoids with C6-C3-C6 skeleton with two aromatic rings A and B and a (dihydro) benzopyran ring C adjacent to A.

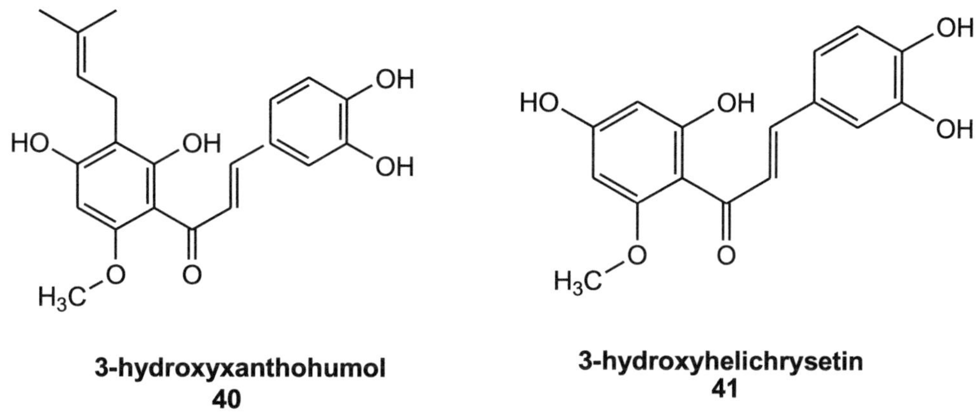

Figure 13.
Prenylated chalcone, 3-hydroxylated derivative of xanthohumol (**40**) *and its nonprenylated analogue 3-hydroxyhelichrysetin* (**41**).

alternating ketones or reduced ketones and methylene groups. Polyketides, owing to their structural diversity and acute toxicity, find applications in medicine, agriculture, and industry. The substitution with prenyl moieties either at a carbon atom of the polyketide nucleus or connection via an ether linkage is a prominent feature in most of these metabolites.

Epoxyphomalin A and B **(42, 43)** (**Figure 14**) are the prenylated polykedtides isolated from marine fungi *Phoma sp* and have strong cytotoxic properties toward six cancer cell lines [26]. Arugosins G and H **(44, 45)** (**Figure 15**) are prenylated polyketides isolated from marine fungus *Emericella nidulans var. acristata*. Arugosin H may be derived from chrysophanol anthrone, which undergoes oxidative cleavage

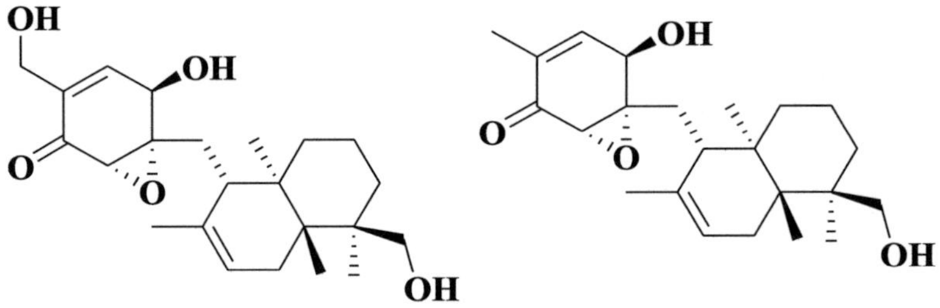

Epoxyphomalin A
42

Epoxyphomalin B
43

Figure 14.
Prenylated polyketides—Epoxyphomalins.

to form the aldehyde function, followed by C-prenylation and hydroxylation. The aldehyde function can be converted to a hemiacetal function, as seen in the tricyclic arugosin G. Arugosins C, D, E **(46, 47, 48)** (**Figure 15**) also occur in Aspergillus spp., whereas arugosin F **(49)** is found in *Ascodesmis sphaerospora* [27].

Prenylated phenyl polyketides named peplidiforones A–D **(50, 51, 52, 53)** (**Figure 16**) are isolated and characterized from *Hypericum peplidifolium*. Unusual carbon skeleton consisting of a furan ring substituted by a 2, 2-dimethylbut-3-enoyl moiety possessed by Peplidiforone C **(52)** is the first example of a prenylated furan derivative isolated from the genus *Hypericum*. The peplidiforones are reported to possess antimicrobial, cytotoxic, antidepressive, antioxidant, and anti-inflammatory effects [28].

Three novel and unusual prenylated polyketides, namely oumarone **(54)**, bissaone **(55)**, and aissatone **(56)** (**Figure 17**), have been isolated from *Harrisonia abyssinica* [29]. Extracts of the bark and the root of this plant exhibit in vitro antiviral, antibacterial, antifungal, and molluscicidal activities [30].

2.4 Terpenoids

The organic compounds derived from the 5-carbon compound isoprene **(57)** and their polymers known as terpenes are collectively called as terpenoids. They are produced by various genera of plants, algae, sponges, and fungi. Terpenoids constitute about 60% of the secondary metabolites produced by plants, known till date. Due to their broad spectrum of medicinal applications, the terpenes have gained significant pharmaceutical value. Without clear distinction, the terms, terpene and terpenoid are usually used interchangeably in the literature. The tree resin terpentine (German: Terpentin; Latin: *Balsamum terebinthinae*) contains a repeating hydrocarbon isoprene unit as a monomer. The etiology of the term "terpene" stems from this tree resin [31]. Generally, in a live plant one can find a terpene, whereas the terpene upon modification with different functional groups and addition or removal of oxidized methyl group makes it a terpenoid. The biological activity of the terpenoids depends on the variation in their structures. The structural unit of a terpene and terpenoid is a five-carbon unit called isoprene **(57)**. These isoprene units are arranged in head-to-head, head-to-tail, and tail-to-tail fashion to give different terpenoids, namely monoterpenes **(58)**, sesquiterpenes **(59)**, diterpenes **(60)**, sesterpenes **(61)**, and triterpenes **(62)** etc. (**Figure 18**) [32].

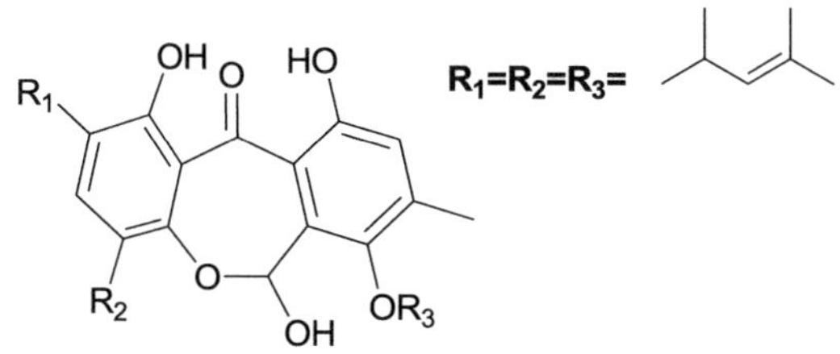

Arugosin G
44

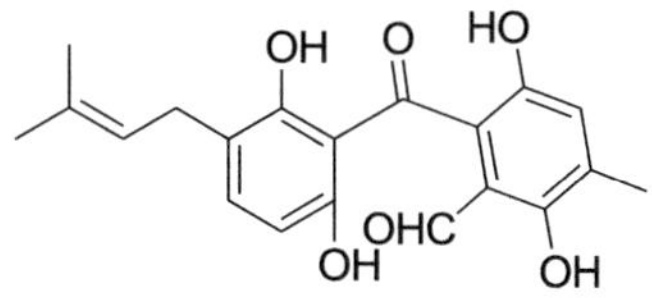

Arugosin H
45

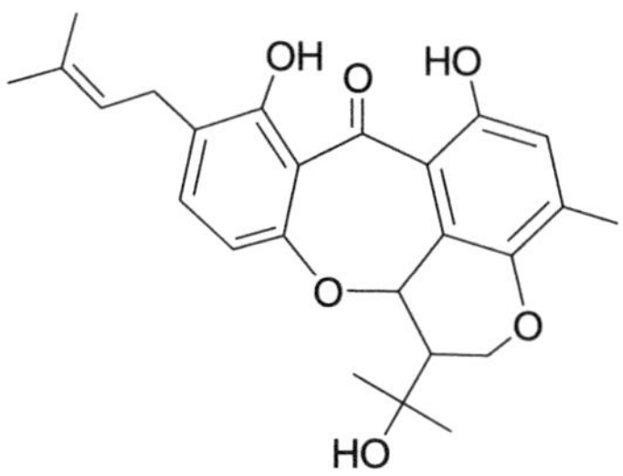

Arugosin C
46

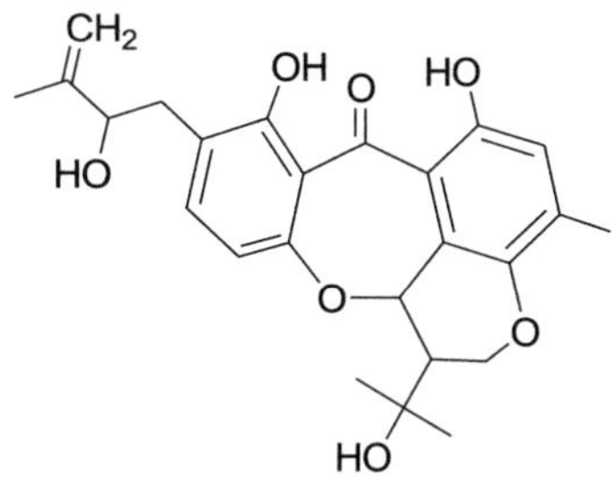

Arugosin D
47

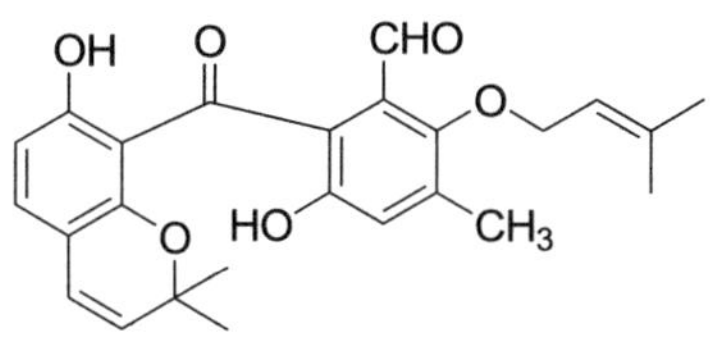

Arugosin E
48

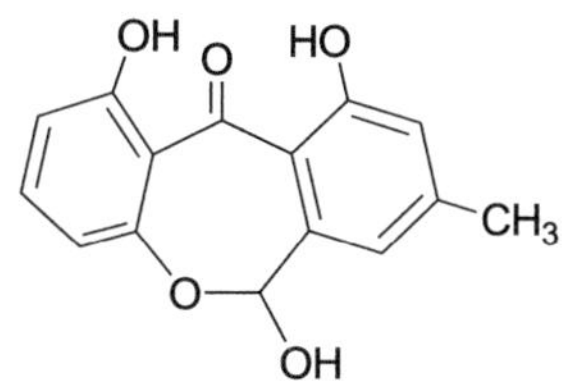

Arugosin F
49

Figure 15.
Prenylated polyketides—Arugosins.

About 1000 prenylated phenolic composite-type terpenoid compounds have been identified to date in plants. The prenylated flavonoids constitute the active components of various medicinal plants. They show sustained biological activities in humans and therefore have been actively investigated as pharmaceuticals [33]. Coumarin derivatives are a group of lactonized phenylpropanoids. The isoprenoid units are not seen in the basic structure of the Furanocoumarins (FCs); therefore the FCs, which are a subgroup of coumarin core with an attached furan ring, are not generally recognized as terpenoid derivatives. However, their furan rings are derived from prenyl chains, followed by the cleavage of a C3 unit to yield the atypical terpenoid derivatives [34].

Two rare antioxidative prenylated terpenoids from loop-root Asiatic mangrove *Rhizophora mucronata* have been isolated. These terpenoids include one new prenylated guaiane sesquiterpenoid **(63)** with an uncommon five-membered lactone ring

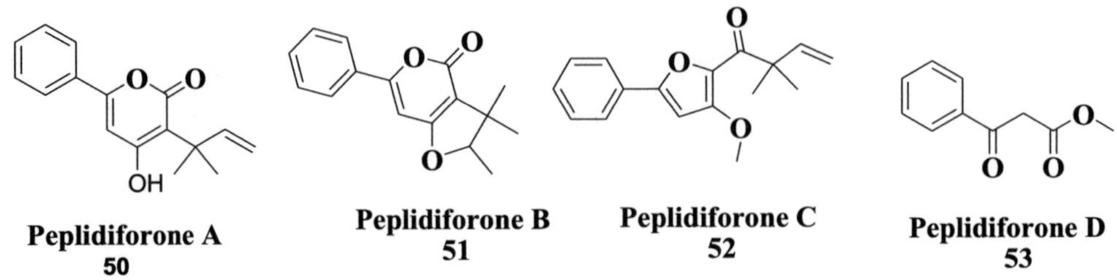

Figure 16.
Prenylated phenyl polyketides—Peplidiforones.

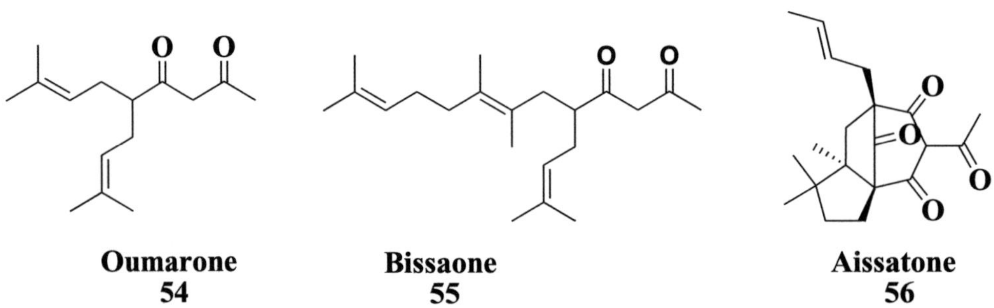

Figure 17.
Unusual prenylated polyketides.

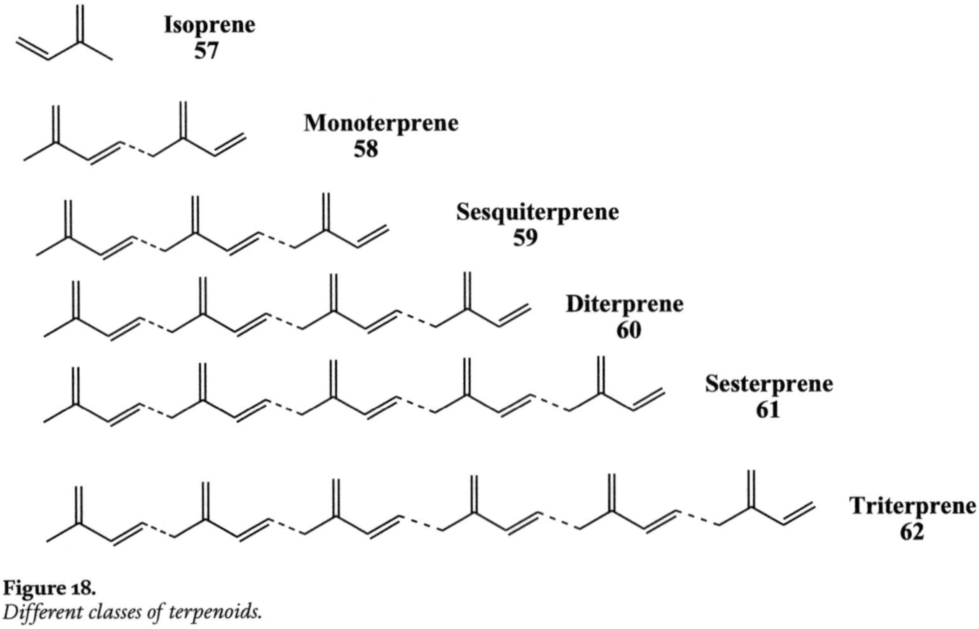

Figure 18.
Different classes of terpenoids.

and prenylated oleanane-type triterpenoid **(64)** (**Figure 19**). These prenylated terpenoids have a potential as lead molecules for use in pharmaceutical and functional food industries [35].

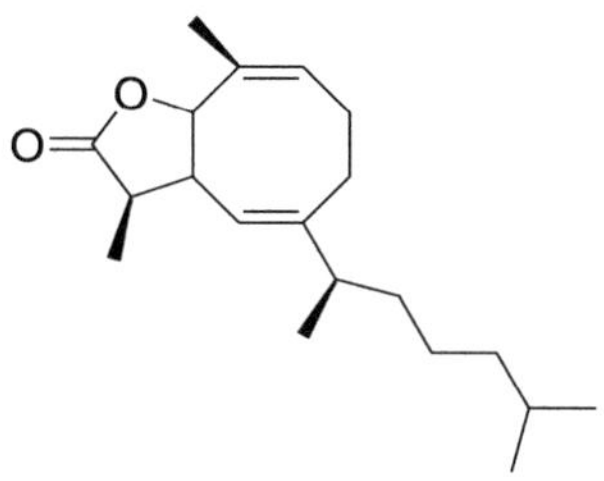

Prenylated guanine sequiterpenoid
63

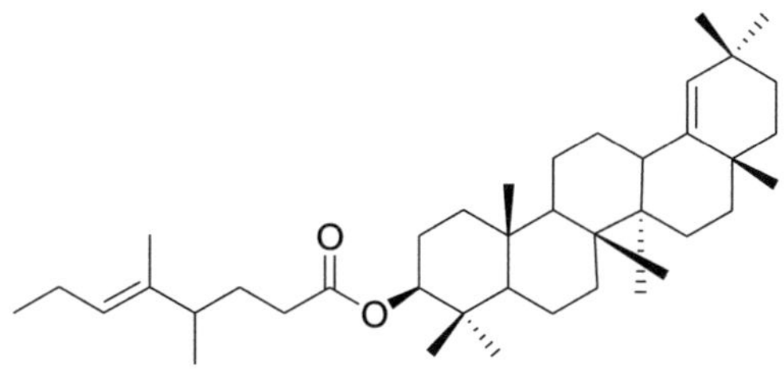

Prenylated oleonone triterpenoid
64

Figure 19.
Rare antioxidative prenylated terpenoids.

3. Prenyltransferases (PTs)

All those enzymes that catalyze the transfer of prenyl groups to a wide variety of acceptors such as proteins, isoprenoid groups, aromatic compounds, etc., are termed as prenyltransferase (PT). PTs are distributed widely in all the living kingdoms and participate in a variety of the metabolic routes [36]. Of late, with increased interest in isoprenoid chemistry, PTs have gained more recognition. The importance of prenylation for the regulation and targeting of bioactive compounds in the cell has been recognized. Among these, the farnesylation of proteins in signal transduction cascades involved in carcinogenesis has been very prominent instance [37].

PTs are unique enzymes in that, apart from creating new C–C bonds, they are also successful in introducing a double bond in the end product. The activation as well as enhancement of the biological activity is generally associated with such features. PTs are peculiar enzymes because they not only create a new C–C bond, a reaction that only some aldolases and lyases have been previously used for [38], but also introduce a double bond in the framework of the final product, a feature that is often associated with the activation or the enhancement of biological properties [39].

The regiospecific/stereoselective chemical synthesis of prenylated aromatic compounds is an arduous task to achieve in good yield, besides the usage of protective groups. But the essential feature in a molecule to exhibit biological activity is its regiospecificity/stereoselectivity. Therefore, an interesting tool for the organic synthesis of biologically active compounds is by the possibility of manipulating enzymatic catalysts such as PTs.

Generally, depending upon the stereochemistry of the resulting products, PTs are divided into two classes, namely cis (or Z) and trans (or E). Dimethylallyltranstransferase is an example of trans-prenyltranferase, whereas dehydrodolichol diphosphate synthase is an example of cis-prenyltransferase.

The transfer of a C5 (dimethylallyl), C10 (geranyl), or C15 (farnesyl) prenyl group derived from the corresponding isoprenyl diphosphate metabolites onto a variety of electron-rich aromatic acceptors is catalyzed by aromatic prenyltransferases. By increasing the affinity for biological membranes and interactions with cellular targets, prenylation provides a higher level of bioactivity compared with the nonprenylated precursor [40]. In a Friedel-Crafts-like reaction, aromatic compounds such as hydroxybenzoic acids and hydroxyphenylketones are prenylated by phenol-oligoprenyldiphosphatase [41]. The role of regiospecific catalysts in widening the horizon of diversity and biological activities of many classes of natural products both *in vivo* and *in vitro* has taken huge interest with recent identification of these enzymes.

4. Conclusion

The prenylated natural compounds exhibit a broad spectrum of interesting molecular, biological, and pharmacological activities. There is a definite consonance between the structure-activity relationship and bioactivities of prenylated natural compounds. The prenyl-moiety increases the chemical diversity and makes the backbone compound more lipophilic, which leads to its high affinity with cell membranes. The prenylation enhances the antibacterial, anti-inflammatory, antioxidant, cytotoxicity, larvicidal as well as estrogenic activities of several natural compounds. Therefore, to fully explore the health-promoting potential, more research is required in the future. Especially the prenyl groups seem to be crucial for the anticancer activity of the natural compounds, possibly leading to enhanced cell membrane targeting and thus increased intracellular activity. Today, cancer prevention is an increasingly important social issue, and the identification and characterization of dietary components or natural products with distinct cancer-preventive qualities and possibly even therapeutic properties, while bearing only low toxicity, are a promising research approach.

Abbreviations

PTases	prenyltransferases
DMATS	dimethylallyl tryptophan synthase
cAMP	cyclic adenosine monophosphate
cGMP	cyclic guanosine monophosphate
PDEs	phosphodiesterases
CNS	central nervous system
MDR	multidrug resistance
Pgp	P-glycoprotein

Author details

Kantharaju Kamanna* and Aravind Kamath
Department of Chemistry, Peptide and Medicinal Chemistry Research Laboratory, Rani Channamma University, Belagavi, Karnataka, India

*Address all correspondence to: kk@rcub.ac.in

Chapter 6

Glycosylation on Spermatozoa, a Promise for the Journey to the Oocyte

Shuangjie Wang, Yadong Li, Aijie Xin, Yang Yang, Sheng-Ce Tao, Yihua Gu and Huijuan Shi

Abstract

Spermatozoa experience a long and tough transit in male and female genital tracts before successful fertilization. Glycosylation helps spermatogenesis, epididymal maturation, passing through cervical mucus, avoiding killing of the female immunologic system, and shaking hands between sperm and egg. Changes in glycosylations along the transit ensure that the right things happen at the right time and place on spermatozoa. Aberrant glycosylations on spermatozoa will negatively affect their fertility. Thus, we developed a lectin array method to examine the glycocalyx of spermatozoa, which will help observe glycosylations occurring on spermatozoa in a normal or abnormal conditions, such as spermatozoa with DEF126 mutation and poor freezability. Intriguingly, binding levels of ABA (*Agaricus bisporus agglutinin*), a lectin marking the inner layer of the glycocalyx, were changed in these subfertile spermatozoa, which indicates that the integrity of glycocalyx is critical for sperm fertility. In this chapter, we reviewed the impacts of glycosylations on sperm fertility, the lectin array method, and its potential application for sperm function assessment.

Keywords: spermatogenesis, sperm maturation, transit through genital tract, glycan profile, lectin microarray

1. Introduction

Before mating the oocytes, mammal spermatozoa go on a long adventure, experiencing a series of complicated events along the male and female reproductive tracts [1]. From spermatogonia, spermatozoa develop in the testis, with dramatic changes in morphology and chromatin structure. However, spermatozoa in testis could not fertilize oocytes since their motility and capabilities of recognizing and binding zona pellucida are underdeveloped.

Spermatozoa then enters into the epididymis, several meters long male reproductive tract, and undergo an additional maturation. During the epididymal maturation, also called post-testicular maturation ranges from 1 to 21 days depending on species [2], spermatozoa acquire not only capacities for progressive motility, acrosome reaction, zona pellucida binding and recognition, and spermatozoon-oocyte fusion [3–7], but also further morphology changes [8–11], such as chromatin condensation, head size decrease, acrosome reshaping, and droplet migration.

After ejaculation, spermatozoa are transferred into the vagina. Passing through the cervix, uterus, and uterotubal junction, spermatozoa stop at a storage reservoir in the oviduct until ovulation. During ovulation, spermatozoa escape the reservoir, complete the capacitation, migrate to the site of mating, and finally achieve insemination. Despite facilitation guiding to the site of fertilization, a successful spermatozoon faces big challenges in the female genital tract, such as selection, which ensures the fittest one to fertilize, and immune defense, which fights evasion of microbial pathogens [12, 13].

2. Glycosylations on mammal spermatozoa and their physiological functions

Along the journey of spermatozoa, events, such as sperm development, maturation, transit, and fertilization, are mediated by cell–cell interaction, and interaction between spermatozoa and the microenvironment [1]. Providing promises to these events, dramatic modifications on sperm surface occur along the transit of spermatozoa [14]. For instance, high sialylation provides a negative charge surrounding the sperm surface, which results in a repulsive interaction against negatively charged mucins for bypassing cervix mucus [15]; a shell, which protects sperm antigens from immune surveillance in the female genital tract [16]; and an anchor, which helps the epithelial cells hold spermatozoa at the oviductal reservoir [17]. However, during capacitation, lipid component changes resulting in a high membrane fluidity for sperm hyperactivation [18], and desialylation occurs on the sperm surface [19, 20], both of which ensure spermatozoa reach the oocyte and fertilize successfully.

2.1 Glycoconjugates on mammal spermatozoa

A dense glycoconjugate coat with a thickness of 20–60 nm called glycocalyx on animal cells [21], is one of the major surface components on spermatozoa, which participate in such modifications and play critical roles in those interactions along the sperm's journey to the oocyte [22, 23]. The glycocalyx is synthesized during spermatogenesis in testis and is remodeled through epididymal maturation, the mixing process of sperms and seminal plasma secreted by the accessory glands during ejaculation and transit through the female genital tract [21]. Two of three types of glycoconjugates are identified in sperm glycocalyx: glycoproteins and glycolipids [24], which are classified corresponding to types of glycan-bearing carriers, proteins, or lipids. Besides those glycoconjugates inserted or anchored in the plasma membrane, others are non-covalently attached to the membrane via inter-molecule or hydrophobic interaction [21].

2.1.1 Glycoproteins

Glycosylation is a ubiquitous posttranslational modification on proteins, more than half of human proteins are deduced as glycoproteins according to their amino acid sequences [25]. By exploiting their covalently linked glycans, membrane or surface glycoproteins play roles in intercellular signaling, cell–cell interaction, and cell adhesion [26]. With nucleoside diphosphate or monophosphate glycans as precursors, glycosyltransferases create a glycosidic bond between glycan and peptide backbone. Three types of covalent linkages exist naturally between glycans and polypeptide backbone: N-glycosylation, O-glycosylation, and GPI glycosylphosphatidylinositol [27].

As the most common posttranslational modification in the endoplasmic reticulum (ER) [28], N-linked glycosylation attaches glycans to the amide nitrogen of asparagine or arginine residues within the consensus amino acid sequences or called "sequons," Asn-X-Thr/Ser, which direct protein folding. In a manner of protein-, cell-, or species-specificity [29], remodeling in Golgi apparatus endows N-glycan structural and functional varieties, which ensure functional complexity of cell surface proteins, such as cell–cell, cell-matrix interaction, or immune responses. N-glycosylation is a prominent posttranslational modification of the protein, about 50% of the proteins have N-glycans [30, 31]. N-glycans are also major carbohydrates on the sperm glycocalyx, which are composed of three types of N-glycans, high mannose, hybrid N-glycans, and complex N-glycans [32]. High mannose N-glycans are essential for Sertoli-germ cell attachment during spermatogenesis, which is a critical cell–cell adhesion for sperm development in testis [33].

Initially in the ER then Golgi apparatus, O-glycosylation covalently links glycans to the oxygen atom at the hydroxyl group of serine, threonine, tyrosine, hydroxylysine, or hydroxyproline residue on the peptide backbone [27]. O-glycosylation has no sequon and core structure, and O-glycan moieties range from monosaccharides to sulfonated polysaccharides. β-defensin 126 (DEFB-126) is a well-known O-linked glycoprotein [34], which is acquired through epididymal maturation and is also an important component of sperm glycocalyx. On the surface of DEFB-126 knock-out sperm, the level of O-glycosylation significantly decreased, and the spermatozoa are male infertility [35].

Glycosylphosphatidylinositol (GPI)-anchor is glycolipid, containing a Mana1–6Mana1–4GlcNa1–6PI motif [36]. Performed by ER-situated transamidase in ER, the carboxyl-terminus of proteins is connected to the ethanolamine moiety of GPI-anchor via an amide bond. Located on the extracellular side of the cytoplasmic membrane, the protein is bound to the cell membrane through GPI-anchoring. Structures and functions of GPI-anchored protein are widely diverse and play important roles in many biological processes. During sperm migration in the epididymis, various GPI-anchored proteins are obtained on the surface under the mediation of epididymosomes in the epididymal fluid. A well-studied GPI-anchored protein is sperm adhesion molecule 1(SPAM1/PH-20). The SPAM1 secreted by the epididymis exists in both epididymal fluid and epididymal exosomes; SPAM1 in the latter can be attached to the epididymal sperm surface by GPI anchoring [37].

2.1.2 Glycolipids

In addition, to proteins, lipid moieties are another type of glycan carrier. Glycosphingolipids are major glycolipids in the cell membrane of animals, which are primarily synthesized in the endoplasmic reticulum, and whose carbohydrate moieties are further modified in the Golgi apparatus [38]. Glycosphingolipids contain one or more carbohydrate residues linked to a specific hydrophobic lipid moiety, sphingoids, or ceramides (N-acylated sphingoid), through a glycosidic bond [39]. According to their glycan structures, glycosphingolipids are classified into the following, namely, ganglio-, isoganglio-, lacto-, neolacto-, lactoganglio-, globo-, isoglobo-, muco-, gala-, neogala-, mollu-, arthro-, schisto- and spirometo-series.

Glycolipids participate in cellular functions in animals, and their biosynthesis and distribution change corresponding to different physiological functions and stages [40]. Glycolipids are essential for many processes of male reproduction [41]. Via manipulated deficiency of galactosyl transferase or sulfotransferase, mouse models with targeted deletion of seminolipid, a sulfoglycolipid, show suspension of spermatogenesis at the stage of meiosis [42]. And, a mouse model with complex ganglioside deficiency resulting from targeted deletion of N-acetylgalactosaminyl transferase is reported to be male infertility [43].

2.2 Physiological functions of sperm glycoconjugates

2.2.1 Spermatogenesis

Occurring in seminiferous tubules of testis, spermatogenesis is a highly regulated series of sperm development, involving in the proliferation of spermatogonia, differentiation, and maturation into spermatozoa. Sertoli-germ cell interaction is vital for spermatogenesis [44], which supports, transports, and protects germ cells. Mouse model studies targeted deletion of N-acetylglucosaminyltransferase-II (GnT-II) and α-mannosidase IIx (MX) suggested that a specific sperm N-glycan, GlcNAc-terminated triantennary, and fucosylated N-glycan, mediates the adhesion of germ cells to Sertoli cells via protein-carbohydrate interaction [45, 46].

Although N-glycans play a role in spermatogenesis has been widely admitted [32, 33], proteins containing, such as, glycans remain little known. Leukocyte differentiation antigen (basigin, BSG, also known as CD147) is an essential protein during the process of spermatogenesis, and embryo implantation [47, 48]. which is expressed in various stages of spermatogenic cells (spermatogonia, spermatocytes, and spermatids), Sertoli cells, Leydig cells, and epididymis [49]. Mature spermatozoa are absent in the testis of a mouse model with targeted deletion of BSG. Mass spectrometry and lectin studies reported that N-glycans containing fucosylated high-mannose moieties with terminated N-acetylglucosamines were identified in BSG [49–51], which is reasonably believed that BSG may participate in Sertoli-germ cell adhesion.

As sulfated glycoconjugates, seminolipids are predominant glycolipids on the plasma membrane of mammalian male germ cells [52]. With a very high cell specificity, seminolipids are synthesized in primary spermatocytes and exist in the subsequent male germ cell types [53]. Seminolipid-deficient mice illustrate male infertility and suspension of spermatogenesis before first meiotic division [42], which might result from failure of uptake of lactate for spermatocytes [54]. *In vitro* studies indicated that Sertoli membrane protein could bind seminolipids, which suggests that seminolipids may take part in Sertoli-germ cell adhesion during spermatogenesis [55].

Gangliosides are sialic acid terminated glycolipids. Complex ganglioside deficient mice showed male infertility, which might result from failure of testosterone transportation in Leydig cells [43]. Due to their reported functions on carbohydrate-dependent cell adhesion [56], gangliosides are also supposed to play roles in Sertoli-spermatocyte interaction [55].

2.2.2 Epididymal maturation

Through the maturation along the epididymal tract, spermatozoa acquire not only fertility, including progressive movement,, but further modifications of morphology (especially, migration of the cytoplasmic droplet), molecular components and cell surface.

During the transit from caput down to cauda epididymis, a significant change in sperm surface is an increase in surface sialylation [56, 57]. Sialic acid is a strongly negatively charged acidic oligosaccharide, sialylation occurs on the terminus of N- or O-glycans. Due to the sialylation on the surface, spermatozoa are covered with a negative charge [58], which could facilitate sperm transit along the genital tracts, shield sperm antigens, and protect spermatozoa from an immune response in the female genital tract [24].

Integrating the sialylated glycoproteins or glycopeptides secreted from epididymal epithelium to the cell surface is a major means to sialylate the sperm surface. Several epididymal sialylated proteins have been reported to be associated with the spermatozoa during epididymal maturation, including proteins D and E, basigin, CD59, fertilin, HE2, HE4 [59], and HE5/CD52 [59, 60].

Secreted from the epididymal epithelium into the epididymal lumen, recent studies showed its significance in the sialylation of sperm surface [61]. CD52 is N-glycosylated mainly in a form of three-antennary oligosaccharides with a terminus of sialylation, which results in a negative charge on CD52. CD52 protects sperm from complement-dependent immune responses, facilitates the process of sperm glycocalyx remodeling, and prevent sperms from agglutination with each other [61, 62]. CD52 is a GPI-anchored protein [61], which is transported and integrated to the sperm surface via epididymosome [63].

Acquisition of membrane lipids from epididymal epithelium also exploits epididymosome [64], relative quantities of cholesterol and sphingomyelin increase during the epididymal maturation. However, other strategies are also applied for the remodeling of sperm surface components.

Originated in the epididymal epithelium and secreted into the lumen, DEFB-126 is also a highly sialylated O-linked glycoprotein, which tightly associates with the surface of human and macaque spermatozoa. It is believed that DEFB-126 is a major component of sperm glycocalyx, and a predominant contributor to the negative charge of sperm surface [34, 35, 65].

The majority of glycosyltransferase activities are found within the epididymal lumen [66]. High activities of both fucosyltransferase and sialyltransferase exist in caput epididymis and decrease gradually at cauda epididymis. It is suggested that further glycosylation may occur on the sperm surface.

2.2.3 Transit along the female genital tract

After ejaculation, spermatozoa will encounter not only selection pressure, which knocks out spermatozoa with poor motility or morphology via retrograde flow in the female genital tract [67], but immune pressures, which eliminates "foreigners," such as microbial pathogens, and even spermatozoa.

Like CD52 mentioned above, CD55, also a GPI-anchored protein, is a type of complement-regulating protein on the plasma membrane of human sperm against immune attack [68]. CD59 is an inhibitory molecule of the membrane attack complex of the complement system, which is also localized in the sperm plasma membrane [69]. These three GPI-anchored proteins with N-glycans protect spermatozoa from immunological attack in the female genital tract [70].

DEFB-126 is an O-linked glycoprotein. Like CD52, CD55, and CD59, DEFB-126 is integrated to sperm surface during epididymal maturation, but in a different way [65]. High sialylation on DEFB-126 not only protects spermatozoa from phagocytosis in the female genital tract, but helps penetrate through cervix mucus. It is supposed that tri- and tetra-antennary glycans with Lewis-X and Lewis-Y sequences on these glycoproteins acquired in epididymis may help escape immune detection [71], and suppress antigen-immune responses [72].

Besides pressures, the female genital tract also conducts facilitation and regulations, which help spermatozoa do the right thing at the right time, right place. Glycodelin is a secretary protein with N-glycans, which is composed of four glycoforms, namely, glycodelin-A (amniotic fluid) [73, 74], glycodelin-F (follicular fluid) [75], glycodelin-C (cumulus matrix) [76], and glycodelin-S (seminal plasma) [77], according to the tissues and fluids where the isoforms are identified. By fucosyltransferase-5 on sperm surface, glycodelin isoforms bind to spermatozoa with a spatiotemporal specificity [78], and perform diverse functions [79], with their different glycan moieties [80].

After ejaculation, spermatozoa bind glycodelin-S on the whole head [75], an abundant protein in seminal plasma secreted from the seminal vesicle, which inhibits albumin-induced cholesterol efflux and prevents sperm capacitation. After cervix penetration, glycodelin-S on spermatozoa is displaced by glycodelin-A, which is secreted from maternal endometrial epithelial cells into uterine and amniotic fluid [80], and covers the sperm acrosomal region. Glycodelin-A inhibits sperm capacitation and protects spermatozoa from immune attack. In the fallopian tube, glycodelin-F binds to spermatozoa in the acrosomal region and inhibits the progesterone-induced acrosome reaction. GdC in the cumulus matrix displaces sperm-bound glycodelin-A and -F during sperm cumulus penetration, covers spermatozoa in the equatorial region, and promotes the zona binding capacity of the spermatozoa.

Glycodelin-A carries high mannose, hybrid, and complex-type structure [73]. Whereas, glycans on glycodelin-S are much different, which are highly fucosylated, and contain a complex-type structure of bi-antennary glycans with Lewis-X and Lewis-Y antennae. Glycodelin-A, -F, -C have a similar topology of carbohydrate moiety, with different degrees of sialylation. Glycodelin-A and glycodelin-F are diversely sialylated, which are immunosuppressive, while non-sialylated glycodelin-S works on inhibition of capacitation, and low-sialylated glycodelin-C as successful singleton insemination [79].

In the female genital tract, spermatozoa also take part in the remodeling of the sperm surface. Ovulation triggers the release of sperm at the oviductal reservoir and sperm capacitation. During their last stage of the journey to the oocyte, spermatozoa carry out a series of desialylation on the sperm surface, including releasing high-sialylated DEFB-126 from spermatozoa to free the latter in the oviductal reservoir [17], the release of two neuraminidases (NEU1 and NEU3) to remove sialic acids on sperm surface [81], and release of sialylated GPI-anchored glycoprotein via raft redistribution due to cholesterol efflux, such as CD52 [82]. Desialylation and raft redistribution joined by glycolipids, such as ganglioside GM1and seminolipids, help present functional surface proteins for cumulus penetration, such as SPAM1 [83], for sperm-zone pullucide recognition and binding, such as dicarbonyl/L-xylose Reductase (P34H) [84], and cysteine-rich secretory protein-1 (CRISP1) [85].

3. Lectin microarray, a "sugar decipher" for sperm surface glycoconjugates

Glycosylation ubiquitously exists in cellular biology, however, it is still a poorly understood posttranslational modification, due to its complexities. One is the vast diversity of glycan structure [86]. Due to carbohydrate structural and linkage features, it is established that all possible branched or linear oligosaccharide isomers yield 1.05×10^{12} structures.

The other is the complexity of glycan biosynthesis, due to the complicated biosynthesis machinery and no preexisting template molecule. Possibly, 1–2% of the genome produces glycan biosynthesis-related proteins, such as glycosyltransferases, glycosidases, and transporters. Carbohydrate biosynthesis does not follow "the central dogma," [87] but several factors instead [88], such as levels, activities, and trafficking of glycosyltransferases, availabilities of carriers and nucleotide sugar substrates, different physiological functions and statuses, and a variety of outer stimuli [89], such as nutrition [89], oxidative stress [90], and so on.

In other words, extreme complexity enables glycans as "the third group of bio-informative molecules," [86] which integrate panoramic information on cell biology [91, 92], including genomes encoding proteins and biosynthesis machinery of lipids or glycans, and interactions between cell and environmental factors. Glycan profiles on cell surfaces help distinguish individuals and physiological statuses. For an instance, α-linked N-acetylgalactosamine (GalNac) and β-linked GalNac on Sda/GM2 is exploited to identify spermatogonial subpopulations in cattle, pigs, and horses [93].

3.1 The lectins and other glycan-binding proteins

Lectins are a group of glycan-binding proteins that selectively bind different glycans with specific structures, which are first described back in the late of the nineteenth century [94]. As more and more plant lectin-glycan pairings have been identified [95], a sophisticated, reliable, and convenient tool set emerged (Summarized in **Table 1**). Although lectins could not acquire detailed structural information, they are capable to tell the differences between glycan profiles on different cell surfaces [96].

Animal lectins are endogenous glycan-binding proteins, which are involved in a variety of biological processes, and could provide more biological function clues for glycans (Summarized in **Table 2**). Intriguingly, summarized in **Table 3**, a group of sperm-egg receptors on spermatozoa are reported to have the capability of glycan-binding, including lectin-like proteins, glycosyltransferases, and glycosidases [132, 133]. We identified a chitobiase, lysosomal di-N-acetylchitobiase (CTBS), on mouse spermatozoa [Data unpublished]. CTBS is expressed on spermatozoa with species specificity and shows the binding ability of Lewis-X, which could inhibit mouse sperm-egg binding [134].

Microbial nontoxic B subunits and developed antibodies against specific carbohydrate structures are also glycan-binding proteins, which are exploited to probe the specific glycans [135, 136]. Cholera toxin subunit B (CTxB) is the nontoxic subunit of CTx, which specially bind to the oligosaccharide moiety of a raft-associated glycosphingolipid, ganglioside GM1, and mediates entry of microbe into the host cells [137]. In male fertility, GM1 localization on human spermatozoa detected by CTxB is used as a diagnostic tool for evaluating sperm response to stimuli for capacitation [138].

3.2 Lectin microarray methods

Originated from protein chips, the technique of lectin microarray is established in 2005 for high throughput analysis of glycans [139]. Compared with mass spectrometry-based approaches, analysis of lectin microarray is simple, convenient, and low

Lectins	Carbohydrate specificity	Sources
AAA	Fucose	*Anguilla Anguilla*
AAL	Fucα1–2,3,4	*Aurentia*
ABA	Galβ1–3GalNAc	*Agaricus bisporus*
ACA	Galβ1–3GalNAc	*Amaranthus caudatus*
AMA	Manα-	*Arum maculatum*
APA	Mannose	*Allium porum*
APP	GalNAcα-, GalNAcβ-	*Aegopodium podagraria*
ASA	Manα1–3	*Allium sativum*
BDA	GalNAcα-, GalNAcβ-	*Bryonia dioica*
BBC	GalNAcα-, GalNAcβ-	*Phaseolus vulgaris sp.*
BPL	Galβ1–3GalNAc	*Bauhinia Purpurea*
CA	Galβ1–4GlcNAc, GalNAcβ1–4GlcNAc	*Colchicum autumnale*
CAA	Complex glycans (GlcNAcβ1–2Manα1–3(GlcNAcβ1–2Manα1–6)Manβ1–4GlcNAcβ1–4GlcNAcβ-)	*Caragana arborescens*
CALSEPA	Mannose, Glucose, Glcα1–4Glc	*Calystegia sepiem*
CCA	Neu5Ac	*Cancer antennarius*
Con A	Manα-, Gluα-	*Canavalia ensiformis*
CPA	Mannose	*Cicer arietinum*
CSA	GalNAcα-	*Cytisus sessilifolius*
DBA	GalNAcα1–3GalNAc, GalNAcα1–3Gal	*Dolichos biflorus*
DSL	Neu5Ac-Gal/GalNAc	*Datura Stramonium*
ECA	Galβ1–4GlcNAc	*Erythrina cristagalli*
EEA	GalNAcβ-	*Euonymus europaeus*
GHA	Galα-, GalNAcα-	*Glechoma hederacea*
GNA	Manα-	*Galanthus nivalis*
GS-I	Galα-, GalNAcα-	*Griffonia simplicifolia*
GS-IA4	GalNAcα-	*Griffonia simplicifolia*
GS-IB4	Galα-	*Griffonia simplicifolia*
GS-II	GlcNAcα-, GlcNAcβ-	*Griffonia simplicifolia*
HAA	GlcNAcα-, GalNAcα-	*Helix aspersa*
HHA	Manα-	*Hippeastrum hybrid*
HMA	GalNAcα-, Fucα-, Neu5Ac	*Homarus americanus*
HPA	GalNAcα-	*Helix pomatia*
IAA	GalNAc	*Iberis amara*
IRA	GalNAcα-, GalNAcβ-	*Iris hybrid*
Jacalin	Galα-, Galβ-, GalNAcα-O link	*Artocarpus integrifolia*
LAA	GlcNAcβ-, GlcNAcβ1–4GlcNAc	*Laburnum alpinum*
LAL	Fucose	*Laburnum anagyroides*
LBA	GalNAcα-, complex glycans (GalNAcα1–3(Fucα1–2)Gal)	*Phaseolus lunatus*

Lectins	Carbohydrate specificity	Sources
LcA	Mannose	*Lens culinaris*
LcH	Complex glycans (Man/GlcNAc core with Fucα1–6)	*Lens culinaris*
LcH A	Manα-, Glcα-, GlcNAc	*Lens culinaris*
LcH B	Manα-, Glcα-, GlcNAc	*Lens culinaris*
LEL (TL)	GlcNAc	*Lycopersicon esculentum*
LFA	Neu5Ac	*Limax flavus*
LPA	Neu5Ac	*Limulus polyphemus*
LTL	Fucα1–2,3,4	*Lotus tetragonolobus*
MAA	Neu5Acα2–3Gal	*Maackia amurensis*
MAL-I	Galβ1–4GlcNAc	*Maackia Amurensis*
MAL-II	Neu5Acα2–3	*Maackia Amurensis*
MNA-G	Galα-, Galβ-	*Morniga G*
MNA-M	Manα-	*Morniga M*
MOA	Galα1–3	*Marasmium oreades*
MPA	Galα-, GalNAcα-	*Maclura pomifera*
NPA	Mannose	*Narcissus pseudo-narcissus*
PSA	L-Fucα1,6GlcNAc	*Pisum sativum*
PHA-E	Complex glycans (Galβ1–4GlcNAcβ1–2(Galβ1–4GlcNAcβ1–6) Man)	*Phaseolus vulgaris*
PHA-L	Complex glycans (Galβ1–4GlcNAcβ1–2Man)	*Phaseolus vulgaris*
PMA	Manα1–3	*Polygonatum mulitiflorum*
PNA	Galβ1,3GalNAc	*Arachis hypogaea*
PSL	Complex glycans (Neu5Acα2–6Galβ1–4 GlcNAc, Neu5Acα2–6Galβ1–4Glc)	*Polyporus squamosus*
PTA Gal	Galactose	*Psophocarpus tetragonolobus*
PTA GalNAc	GalNAc	*Psophocarpus tetragonolobus*
PWM (PWA)	GlcNAcβ1–4GlcNAc	*Phytolacca americana*
RCA-I	Galβ-	*Ricinus communis*
RCA-II	Galβ-, GalNAcβ-, Galβ1–4Glc (Lactose)	*Ricinus communis*
SBA	GalNAcα/β-	*Soybean*
SHA	GalNAc	*Salvia horminum*
SJA	GalNAcβ-	*Sophora japonica*
SNA	Neu5Acα2–6Gal/GalNAc	*Sambucus nigra*
SNA-I	Neu5Acα2–6Galβ1–4GlcNAc, Neu5Acα2–6 Galβ1–4Glc	*Sambucus nigra*
SNA-II	Galβ-, GalNAcβ-	*Sambucus nigra*
SSA	GalNAc-O link	*Salvia sclarea*
STL (PL)	GlcNAc, Neu5Ac	*Solanum Tuberosum*
Succinyl-Con A	Mannose	*C. ensiformis*
TKA	Galβ-, Galβ1–4Glc (Lactose)	*Trichosanthes kirilowii*

Lectins	Carbohydrate specificity	Sources
TL	α-GalNAc, β-GalNAc, GalNAc, Galatose, Fucose	*Tulipa sp.*
UDA	GalNAcβ-	*Urtica dioica*
UEA-I	Fucα-, Fucα1–2Galβ1–4GlcNAc	*Ulex europaeus*
UEA-II	GlcNAcβ-	*Ulex europaeus*
VAA	Galβ-	*Viscum album*
VFA	Manα-	*Vicia fava*
VGA	Galβ1–3GalNAc	*Vicia graminea*
VRA	Galα-,	*Vigna radiata*
VVA	GalNAcα-, GalNAcα1–3Gal	*Vicia villosa*
VVA Man	Mannose	*Vicia villosa*
WFA	GalNAcα/β-	*Wisteria floribunda*
WGA	GlcNAcβ-	*Triticum vulgaris*

The lectin carbohydrate specificities were summarized from: 1, the Consortium for Functional Glycomics (http://functionalglycomics.org); 2, Product information from EY or Vector Laboratories, Inc.

Table 1.
Incomplete list of commercial plant lectins.

time-consuming [140]. Due to its non-quantitation and incomplete determination of glycan structures, lectin microarray is more appropriate for analyzing differences between glycan profiles.

We first established the methodology of lectin microarray for the glycan profiling of sperm surface [141]. The major process is summarized in **Figure 1**. Briefly, the process includes the following steps:

1. Pre-treatment of lectin microarray and propidium iodide (PI)-labeling of fixed spermatozoa

2. Incubation of spermatozoa with the microarray

3. Rinsing the microarray to remove the excess or unbound spermatozoa

4. Air-drying the microarray, and record the signals on the chip

5. Data analysis

6. Verification of the results by using flow cytometry with fluorescence-labeled lectins.

3.3 Sugar codes on spermatozoa with lectin microarray

By using a chip with 91 plant lectins following a previous study [142], we compared surface glycan profiles of five mammalian spermatozoa, human, boar, bull, goat, and rabbit [143]. Roughly, 50 lectins were observed to bind to spermatozoa of 5

Lectin types	Sub-types	Common structures	Members	Carbohydrate specificity	Reference
R-Type	The mannose receptor family	Type I trans-membrane glycoproteins containing a single fibronectin type II domain, multiple C-type lectin domains (CTLDs), and an amino-terminal cysteine-rich domain.	The mannose receptor	C-type domain: acetylglucosamine, mannose, and fucose; C-type domain: sulfated glycans containing 3-O-sulfated galactose, 3-O-sulfated Lex, and 3-O-sulfated Lea;	[97]
			The phospholipase A2 receptor	PLA2 neurotoxins	[98]
			DEC-205	Phosphorothioated cytosine–guanosine (CpG) oligonucleotides	[99]
			Endo180	N-acetyl-glucosamine, mannose, and fucose	[100]
	UDP-GalNAc: polypeptide α-N-acetylgalactosaminyl-transferases	Type II transmembrane proteins containing an amino-terminal catalytic domain and a carboxy-terminal R-type domain	ppGalNAcT1–20	UDP-GalNAc	[101]
L-Type	Protein quality control and sorting related	Type I membrane protein with a lumenal portion containing a Ca^{++}-binding domain a proline-rich long hairpin loop called the P domain, and a L-type lectin domain.	Calnexin (CNX) and calreticulin (CRT, missing the cytoplasmic and transmembrane regions)	Misfolded or aggregated glycoproteins	[102, 103]
		Dilysine/diphenylalanine KKFF retention/retrieval motif in cytoplasmic carboxyl terminus	ERGIC-53 and related proteins ERGL, VIP36, and VIPL	Oligomannose-type glycans	[104]
	Others	Pentameric proteins containing L-type lectin folds	Pentraxins and related proteins	Phosphocholine residues on polysaccharides and on phospholipids; carbohydrate derivatives on bacterial polysaccharides	[105]
		Containing laminin G domain-like (LG) modules	Laminins	Heparin. sulfatide, and α-dystroglycan (α-DG)	[106]

P-Type	Mannose 6 -phosphate receptor	Type I trans-membrane glycoproteins containing large extracytoplasmic domains, transmembrane regions, and carboxy-terminal cytoplasmic domains.	Cation-dependent mannose 6-phosphate receptor	N-glycans containing mannose 6-phosphate	[107, 108]
			Cation-independent mannose 6-phosphate receptor		
I-Type	Sialic acid-binding immunoglobulin-like lectins (Siglecs)	Type-1 membrane proteins containing a Sia-binding domain (a V-set domain at amino terminus), and varying numbers of C2-set Ig domains that act as spacers, between Sia-binding site and plasma membrane	Subgroup1: Siglec-1(Sn), Siglec-2 (CD22), Siglec-4 (MAG), and Siglec-15; (Based on sequence similarity)	Sialic acids	[109–112]
			Subgroup2: CD33 related Siglecs. (Siglec-3, -5 to -11, -14 and -16, based on sequence similarity)		
	Others	Immunoglobulin superfamily (IgSF) members other than Siglecs containing a V-set domain.	Paired immunoglobulin-like type-2 receptors (PILRs)	Mucin-like O-glycosylated membrane proteins	[113, 114]
			Platelet endothelial cell adhesion molecule (PECAM)-1	α2–6-linked sialic acids	[115, 116]
			Neural cell adhesion molecule (NCAM) and basigin (CD147)	Oligomannose-type glycans	[117]
			Intercellular adhesion molecule (ICAM)-1	Hyaluronan	[118]

C-Type	The Ashwell-Morell receptor	Transmembrane heteroligomeric glycoprotein complex composed of ASGPR1 (HL-1) and ASGPR2 (HL-2) subunits		Exposed β-linked galactose residues on asialoglycoproteins	[119, 120]
	Collectins	Transmembrane proteins containing an N-terminal triple-helical collagenous region, an α-helical coiled-coil trimerizing neck region, and a C-terminal C-type lectin domain.	Proteins containing 185-Glu-Pro-Asn	Mannose-like sugars	[121]
			Proteins containing 185-Gln-Pro-Asp	Galactose-like sugars	
	Myeloid C-type lectins	Transmembrane proteins containing immunoreceptor tyrosine-based activation motif (ITAM) motifs in their cytoplasmic portion and extracellualr C-type lectin domains	Dectin-1 cluster, including dectin-1, CLEC-1, CLEC-2, LOX-1, CLEC12b, and CLEC9a	Variate of glycans. Dectin-1 bings β-glucans, which are polymers of a backbone of β1–3-linked glucose and side chains of β1–6-linked glucose	[122, 123]
			Dectin-2 cluster including dectin-2, blood DC antigen 2 (BDCA-2), DC immunoactivating receptor (DCAR), DC immunoreceptor (DCIR), CTL superfamily 8 (CLECSF8), and MINCLE (CLEC4E)	Variate of glycans. Dectin-2 bings α-mannans, which are polymers of α1–6-linked mannose with α1–2-linked mannose side chains	
	Selectins	Type-1 transmembrane glycoproteins, containing an N-terminal C-type lectin domain followed by a consensus epidermal growth factor (EGF)-like domain, a series of consensus repeats, and a transmembrane domain with a short intracellular tail	P-selectin on platelets	P-ctin glycoprotein ligand-1 (PSGL-1), mucins containing highly clustered O-glycans bearing SLex antigens and sulfate esters, SLex/a rich sulfated mucins	[124–128]
			E-selectin on endothelial cells	PSGL-1, other glycoproteins that express the SLex antigen on either N- or O-glycans, long-chain glycosphingolipids expressing the SLex antigen	
			L-selectin on leukocytes	PSGL-1, ligands containing sulfated glycans, such as 6-sulfo-SLex on both core-2 O-glycans and on extended core-1 O-glycans	

	Other proteins with C-type lectin domains		A number of proteins with CTLDs have been identified in the pancreas and kidney	Unknown	
Galectins	Proto-typical Galectins	Containing a carbohydrate-recognition domain (CRD)	Galectin-1, -2, -7, -10, -11, -13, -14, and -15	Glycans with terminal β-Gal residues	[129, 130]
	Chimera-type Galectins	Containing a single CRD and an amino terminus with high ratio of proline, glycine, and tyrosine residues	Galectin-3	Glycans with repeating [-3Galβ1–4GlcNAcβ-]n or poly-N-acetyllactosamine sequences	
	Tandem-repeat Galectins	Containing two CRDs connected by a peptide linker	Galectin-4, -8, -9, and -12	α2–3-sialylated glycans, blood group A determinant on a LacNAc core	
Glycosaminoglycan (GAG)-binding proteins		Do not have common folds	Various GAG-binding proteins	Glycosaminoglycan (GAG)	[131]

Classification from Varki A, Cummings RD, Esko JD, et al., editors. Essentials of Glycobiology [Internet]. 3rd edition. Cold Spring Harbor (NY): Cold Spring Harbor Laboratory Press; 2015–2017. Available from: https://www.ncbi.nlm.nih.gov/books/NBK310274/

Table 2.
Major human carbohydrate-binding proteins.

Species	Sperm receptor	Reference
Mouse	β-1,4-galactosyltransferase (β-1,4-GalT)	[132]
	α-D-mannosidase	[132]
	sulphoglycolipid immobilizing protein (SLIP1)	[132]
	56 kDa galactose binding protein (sp56)	[132]
	α-fucosidase	[133]
	fructosyltransferase	[133]
	di-N-acetylchitobiase (CTBS)	Data unpublished
Rat	α-D-mannosidase	[132]
	galactose receptor	[132]
Human	mannose-binding protein	[132]
	α-D-mannosidase	[132]
	β-1,4-galactosyltransferase (β-1,4-GalT)	[132]
	galactose lectin	[132]
	selectin-like molecule	[132]
	PH-20	[133]
	α-fucosidase	[133]
	fucosyltransferase-5 (FUT5)	[78]
	dicarbonyl/L-xylulose reductase (DCXR)	[133]

Table 3.
Sperm surface glycan-binding receptors.

all mammalian species, which illustrates a diversity of glycan structures on spermatozoa. Although lectin binding profiles showed a similarity among the five species, a subtle difference was discovered. SBA, HPA, and VVL (binding to GalNac) and GSL I (recognizing galactose) showed strong binding to spermatozoa of boar, bull, goat, and rabbit, while MAL II, PHA-L, PHA-E + L, PHA-E and SNA bound to human spermatozoa strongly. The results suggested that complex glycan structures (recognized by MAL II, PHA-L, PHA-E + L, and PHA-E) and terminal sialic acids (recognized by SNA) are richer in human spermatozoa.

DEFB-126 is an O-linked glycoprotein, which is a major component of human sperm glycocalyx. It is postulated that DEF-126 carries terminal sialylation that is critical for sperm penetration through cervix mucus. We compared glycan profiles between human DEFB-126 mutant and wild-type spermatozoa with lectin microarray analysis [143]. Intriguingly, compared with wild-type spermatozoa, binding of ABA and Jacalin (both are O-linked glycosylation specific lectins) decreased dramatically, while binding of SNA (a lectin recognizing sialic acids deleted) did not change on DEFB-126 mutant spermatozoa. It is suggested that deficiency of DEFB-126 will cause a decrease of O-linked glycans, while the level of sialylation around spermatozoa does not change, which indicates that a more complicated mechanism may happen for subfertility resulting from DEFB-126 mutation.

We also studied the impacts of cryopreservation on sperm fertility [144]. Decreased MAA (a sialic acid-specific lectin) and increased ABA (O-linked glycosylation specific lectin, generally recognizing the inner layer of glycocalyx) were discovered. Which suggested that cryopreservation may destroy the outermost layer

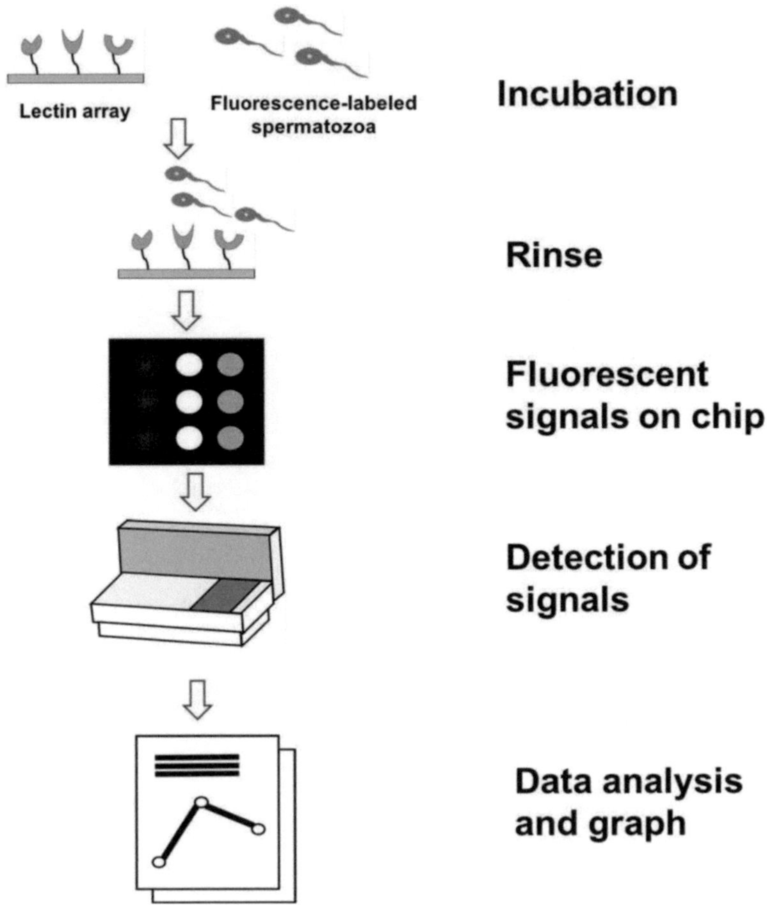

Figure 1.
Scheme of lectin microarray analysis on spermatozoa.

of sialylation on sperm glycocalyx. Furthermore, we found a weak ABA binding to cryopreservation tolerant human spermatozoa, which indicates the importance of the integrity of sperm glycocalyx on cryopreservation tolerance [145].

Due to the significance of glycan structural specificity and biological function, we utilized 60 human lectins and lectin-like proteins for microarray analysis of human spermatozoa [146]. Strong bindings were observed for 5 lectins, galectin-1, 7, 8, GalNAc-T6, and ERGIC-53 (LMAN1). Among them, galectin-8 has definite sperm physiology, which could significantly enhance the acrosome reaction.

4. Conclusion

In modern society, concerns are rising on male fertility. About 10–15% of couples worldwide suffer from infertility, of which male factors account for 50% [147]. Although spermatogenesis is a complicated process, involving in more than 2300 genes that are regulated tempo-spatially to develop spermatogonia to spermatozoa [148], only 30% of male infertility are associated with genetic abnormalities [149, 150]. Furthermore, about 50% of infertility are idiopathic [151, 152], which is a long-term unsolved andrological question, resulting from many possible factors, such as environmental factors, lifestyle, and so on [153]. Thus, a comprehensive andrological assessment is required for a male infertility work-up, including a detailed history analysis, physical examination, well-established semen analysis, endocrine assessment, DNA and epigenetic deficiencies, and so on [154].

The specific molecular mechanism of sperm development, maturation, capacitation, and fertilization could provide clues to solve male reproductive problems. As discussed above, glycan profiling covers panoramic information about genetic background, previous and current environmental impacts, and biological functions of spermatozoa [92]. Due to the universality of glycosylation modifications along the journey of spermatozoa, it is of great significance to study the physiological function of glycosylation on spermatozoa, which could reveal the targets for diagnosis and treatment of male infertility. As a post-genome technique platform, lectin microarray is capable to decode the "sugar code," the glycoconjugate characteristics of cells [139, 155], and its research on spermatozoa can provide new ideas for diagnosis and management of male reproductive system diseases and infertility.

Acknowledgements

We would like to express our special thanks to our all family members and friends who supported us a lot during this harsh lockdown time. We are really thankful for them.

Author contributions

SJW, YDL, AJX, and YY wrote the manuscript; SCT conceived the part of lectin microarray; YHG and HJS are joint corresponding authors.

Conflict of interest

The authors declare no conflict of interest.

Abbreviations

BSG	Leukocyte differentiation antigen basigin also known as CD147
CRISP1	Cysteine-rich secretory protein 1
CTBS	Lysosomal di-N-acetylchitobiase
CTx	Cholera toxin
CTxB	Cholera toxin subunit B
DEFB126	β-defensin 126
ER	Endoplasmic reticulum
GalNac	N-acetylgalactosamine
GalNAc-T6	Polypeptide N-acetylgalactosaminyltransferase 6
GM2	Gangioside GM2
GPI	Glycosylphosphatidylinositol
GnT-II	N-acetylglucosaminyltransferase-II

LMAN1	Lectin mannose-binding 1, endoplasmic reticulum-Golgi intermediate compartment protein 53
MX	α-mannosidase IIx
NEU	Neuraminidases
P34H	Dicarbonyl/L-xylose Reductase
PI	Propidium Iodide
Sda	Sid blood group antigen
SPAM1/PH-20	Sperm adhesion molecule 1

Abbreviates of plant lectins are listed in **Table 1**.

Author details

Shuangjie Wang[1], Yadong Li[2], Aijie Xin[1], Yang Yang[1], Sheng-Ce Tao[3], Yihua Gu[1*] and Huijuan Shi[1]

1 Shanghai Institute for Biomedical and Pharmaceutical Technologies, Shanghai, China

2 Obstetrics and Gynecology Hospital of Fudan University, Shanghai, China

3 Shanghai Jiaotong University, Shanghai, China

*Address all correspondence to: gu_yihua@hotmail.com

References

[1] Yanagimachi R. Mammalian fertilization. The Physiology of Reproduction. 1994;**1**:189-317. DOI: 1571135649640364928

[2] Rowley MJ, Teshima F, Heller CG. Duration of transit of spermatozoa through the human male ductular system. Fertility and Sterility. 1970;**21**:390-396. DOI: S0015028216375021

[3] Lakoski KA, Carron CP, Cabot CL, et al. Epididymal maturation and the acrosome reaction in mouse sperm: Response to zona pellucida develops coincident with modification of M42 antigen. Biology of Reproduction. 1988;**38**:221-233. DOI: 10.1095/biolreprod38.1.221

[4] Yeung CH, Perez-Sanchez F, Soler C, et al. Maturation of human spermatozoa (from selected epididymides of prostatic carcinoma patients) with respect to their morphology and ability to undergo the acrosome reaction. Human Reproduction Update. 1997;**3**:205-213. DOI: 10.1093/humupd/3.3.205

[5] Moore HDM, Hartman TD, Pryor JP. Development of the oocyte penetrating capacity of spermatozoa in the human epididymis. International Journal of Andrology. 1983;**6**:310-318. DOI: 10.1111/j.1365-2605.1983.tb00545.x

[6] Harayama H, Kusunoki H, Kato S. Capacity of rete testicular and cauda epididymal boar spermatozoa to undergo the acrosome reaction and subsequent fusion with egg plasma membrane. Molecular Reproduction and Development. 1993;**35**:62-68. DOI: 10.1002/mrd.1080350111

[7] Avella MA, Xiong B, Dean J. The molecular basis of gamete recognition in mice and humans. Molecular Human Reproduction. 2013;**19**:279-289. DOI: 10.1093/molehr/gat004

[8] Haidl G, Badura B, Schill WB. Function of human epididymal spermatozoa. Journal of Andrology. 1994;**15**(Suppl):23S-27S. DOI: 10.1002/j.1939-4640.1994.tb01699.x

[9] Cooper TG, Yeung CH. Acquisition of volume regulatory response of sperm upon maturation in the epididymis and the role of the cytoplasmic droplet. Microscopy Research and Technique. 2003;**61**:28-38. DOI: 10.1002/jemt.10314

[10] Toshimori K. Biology of spermatozoa maturation: An overview with an introduction to this issue. Microscopy Research and Technique. 2003;**61**:1-6. DOI: 10.1002/jemt.10311

[11] Cooper TG. The epididymis, cytoplasmic droplets and male fertility. Asian Journal of Andrology. 2011;**13**:130-138. DOI: 10.1038/aja.2010.97

[12] Suarez SS. Mammalian sperm interactions with the female reproductive tract. Cell and Tissue Research. 2016;**363**(1):185-194. DOI: 10.1007/s00441-015-2244-2

[13] Miller DJ. Review: The epic journey of sperm through the female reproductive tract. Animal. 2018;**12**(s1):s110-s120. DOI: 10.1017/S1751731118000526

[14] Rickard JP, Pool KR, Druart X, et al. The fate of spermatozoa in the female reproductive tract: A comparative review. Theriogenology. 2019;**137**:104-112. DOI: 10.1016/j.theriogenology.2019.05.044

[15] Tollner TL, Yudin AI, Treece CA, et al. Macaque sperm coating protein DEFB126 facilitates sperm penetration of cervical mucus. Human Reproduction. 2008;**23**(11):2523-2534. DOI: 10.1093/humrep/den276

[16] Czuppon AB. Biochemical characterization of a human spermatozoal sialoglycoprotein with respect to antigenicity masking by its sialic acid moieties. Biochemistry International. 1984;**8**:9-18 PMID: 6206868

[17] Tollner TL, Yudin AI, Tarantal AF, et al. Beta-defensin 126 on the surface of macaque sperm mediates attachment of sperm to oviductal epithelia. Biology of Reproduction. 2008;**78**:400-412. DOI: 10.1095/biolreprod.107.064071

[18] Leahy T, Gadella BM. New insights into the regulation of cholesterol efflux from the sperm membrane. Asian Journal of Andrology. 2015;**17**(4):561-567. DOI: 10.4103/1008-682X.153309

[19] Iqbal N, Hunter AG. Comparison of various bovine sperm capacitation systems for their ability to alter the net negative surface charge of spermatozoa. Journal of Dairy Science. 1995;**78**:84-90. DOI: 10.3168/jds.S0022-0302(95)76619-X

[20] Iqbal N, Hunter AG. Comparison of bovine sperm capacitation systems for ability of sperm to penetrate zona-free hamster oocytes and bovine oocytes matured in vitro. Journal of Dairy Science. 1995;**78**:77-83. DOI: 10.3168/jds.S0022-0302(95)76618-8

[21] Schröter S, Osterhoff C, McArdle W, Ivell R. The glycocalyx of the sperm surface. Human Reproduction Update. 1999;**5**(4):302-313. DOI: 10.1093/humupd/5.4.302

[22] Diekman A. Glycoconjugates in sperm function and gamete interactions: How much sugar does it take to sweet-talk the egg? CMLS. Cellular and Molecular Life Sciences. 2003;**60**:298-308. DOI: 10.1007/s000180300025

[23] Cooper TG. Interactions between epididymal secretions and spermatozoa. Journal of Reproduction and Fertility. Supplement. 1998;**53**:119-136 PMID: 10645272

[24] Tecle E, Gagneux P. Sugar-coated sperm: Unraveling the functions of the mammalian sperm glycocalyx. Molecular Reproduction and Development. 2015;**82**:635-650. DOI: 10.1002/mrd.22500

[25] Apweiler R, Hermjakob H, Sharon N. On the frequency of protein glycosylation, as deduced from analysis of the SWISS-PROT database. Biochimica et Biophysica Acta. 1999;**1473**:4-8. DOI: 10.1016/s0304-4165(99)00165-8

[26] Colley KJ, Varki A, Kinoshita T. Cellular organization of glycosylation. In: Ajit V, Richard DC, editors. Essentials of Glycobiology. 3rd ed. Cold Spring Harbor: Cold Springs Harbor Laboratory Press; 2015. pp. 41-49. DOI: 10.1101/glycobiology.3e.004

[27] Spiro RG. Protein glycosylation: Nature, distribution, enzymatic formation, and disease implications of glycopeptide bonds. Glycobiology. 2002;**12**(4):43R-56R. DOI: 10.1093/glycob/12.4.43r

[28] Yan A, Lennarz WJ. Unraveling the mechanism of protein N-glycosylation. The Journal of Biological Chemistry. 2005;**280**(5):3121-3124. DOI: 10.1074/jbc.R400036200

[29] Aebi M. N-linked protein glycosylation in the ER. Biochimica

et Biophysica Acta. 2013;**1833**:2430-2437. DOI: 10.1016/j.bbamcr.2013.04.001

[30] Varki A, Lowe J. Biological roles of glycans. In: Ajit V, Richard DC, editors. Essentials of Glycobiology. 2nd ed. Vol. 2009. New York: Cold Spring Harbor Laboratory Press; 2009. pp. 75-88 PMID: 20301233

[31] Dennis JW, Nabi IR, Demetriou M. Metabolism, cell surface organization, and disease. Cell. 2009;**139**:1229-1241. DOI: 10.1016/j.cell.2009.12.008

[32] Pang PC, Tissot B, Drobnis EZ, et al. Expression of bisecting type and Lewis-x/Lewis-y terminated N-glycans on human sperm. The Journal of Biological Chemistry. 2007;**282**: 36593-36602. DOI: 10.1074/jbc. M705134200

[33] Huang HH, Stanley P. A testis-specific regulator of complex and hybrid N-glycan synthesis. The Journal of Cell Biology. 2010;**190**:893-910. DOI: 10.1083/jcb.201004102

[34] Yudin AI, Treece CA, Tollner TL, et al. The carbohydrate structure of DEFB126, the major component of the cynomolgus macaque sperm plasma membrane glycocalyx. The Journal of Membrane Biology. 2005;**207**:119-129. DOI: 10.1007/s00232-005-0806-z

[35] Tollner TL, Venners SA, Hollox EJ, et al. A common mutation in the defensin DEFB126 causes impaired sperm function and subfertility. Science Translational Medicine. 2011;**3**:92ra65. DOI: 10.1126/scitranslmed.3002289

[36] Nosjean O, Briolay A, Roux B. Mammalian GPI proteins: Sorting, membrane residence and functions. Biochimica et Biophysica Acta. 1997;**1331**:153-186. DOI: 10.1016/s0304-4157(97)00005-1

[37] Griffiths GS, Galileo DS, Reese K, Martin-Deleon PA. Investigating the role of murine epididymosomes and uterosomes in GPI-linked protein transfer to sperm using SPAM1 as a model. Molecular Reproduction and Development. 2008;**75**(11):1627-1636. DOI: 10.1002/mrd.20907

[38] Maccioni HJ. Glycosylation of glycolipids in the Golgi complex. Journal of Neurochemistry. 2007;**103**(Suppl 1):81-90. DOI: 10.1111/j.1471-4159.2007.04717.x

[39] Yu RK, Tsai YT, Ariga T, Yanagisawa M. Structures, biosynthesis, and functions of gangliosides—An overview. Journal of Oleo Science. 2011;**60**(10):537-544. DOI: 10.5650/jos.60.537

[40] Iwamori M. A new turning point in glycosphingolipid research. Human Cell. 2005;**18**(3):117-133. DOI: 10.1111/j.1749-0774.2005.tb00002.x

[41] Iwamori M, Adachi S, Lin B, et al. Spermatogenesis-associated changes of fucosylated glycolipids in murine testis. Human Cell. 2020;**33**(1):23-28. DOI: 10.1007/s13577-019-00304-x

[42] Honke K, Hirahara Y, Dupree J, Suzuki K, Popko B, Fukushima K, et al. Paranodal junction formation and spermatogenesis require sulfoglycolipids. Proceedings of the National Academy of Sciences of the United States of America. 2002;**99**(7):4227-4232. DOI: 10.1073/pnas.032068299

[43] Takamiya K, Yamamoto A, Furukawa K, et al. Complex gangliosides are essential in spermatogenesis of mice: Possible roles in the transport of testosterone. Proceedings of the National Academy of Sciences of the United States of America. 1998;**95**(21):12147-12152. DOI: 10.1073/pnas.95.21.12147

[44] Mruk DD, Cheng CY. Sertoli-Sertoli and Sertoli-germ cell interactions and their significance in germ cell movement in the seminiferous epithelium during spermatogenesis. Endocrine Reviews. 2004;**25**:747-806. DOI: 10.1210/er.2003-0022

[45] Wang Y, Tan J, Sutton-Smith M, et al. Modeling human congenital disorder of glycosylation type IIa in the mouse: Conservation of asparagine-linked glycan-dependent functions in mammalian physiology and insights into disease pathogenesis. Glycobiology. 2001;**11**:1051-1070. DOI: 10.1093/glycob/11.12.1051

[46] Akama TO, Nakagawa H, Sugihara K, et al. Germ cell survival through carbohydrate-mediated interaction with Sertoli cells. Science. 2002;**295**:124-127. DOI: 10.1126/science.1065570

[47] Igakura T, Kadomatsu K, Kaname T, et al. A null mutation in basigin, an immunoglobulin superfamily member, indicates its important roles in peri-implantation development and spermatogenesis. Developmental Biology. 1998;**194**:152-165. DOI: 10.1006/dbio.1997.8819

[48] Chen L, Bi J, Nakai M, et al. Expression of basigin in reproductive tissues of estrogen receptor-α or -β null mice. Reproduction. 2010;**139**:1057-1066. DOI: 10.1530/REP-10-0069

[49] Bi J, Li Y, Sun F, et al. Basigin null mutant male mice are sterile and exhibit impaired interactions between germ cells and Sertoli cells. Developmental Biology. 2013;**380**:145-156. DOI: 10.1016/j.ydbio.2013.05.023

[50] Tang W, Chang SB, Hemler ME. Links between CD147 function, glycosylation, and caveolin-1. Molecular Biology of the Cell. 2004;**15**:4043-4050. DOI: 10.1091/mbc.e04-05-0402

[51] Loong JH, Wong TL, Tong M, Sharma R, Zhou L, Ng KY, et al. Glucose deprivation-induced aberrant FUT1-mediated fucosylation drives cancer stemness in hepatocellular carcinoma. The Journal of Clinical Investigation. 2021;**131**(11):e143377. DOI: 10.1172/JCI143377

[52] Vos JP, Lopescardozo M, Gadella BM. Metabolic and functional-aspects of sulfogalactolipids. Biochimica et Biophysica Acta (BBA)—Lipids and Lipid Metabolism. 1994;**1211**:125-149. DOI: 10.1016/0005-2760(94)90262-3

[53] Tanphaichitr N, Bou Khalil M, Weerachatyanukul W, et al, Physiological and biophysical properties of male germ cell sulfogalactosylglycerolipid, in: De Vriese S (Ed.), Lipid Metabolism and Male Fertility, AOCS Press, Champaign, IL, 2003, pp. 125-148. DOI:10.1201/9781439822234.ch11

[54] Honke K. Biosynthesis and biological function of sulfoglycolipids. Proceedings of the Japan Academy. Series B, Physical and Biological Sciences. 2013;**89**(4): 129-138. DOI: 10.2183/pjab.89.129

[55] Tanphaichitr N et al. Properties, metabolism and roles of sulfogalactosylglycerolipid in male reproduction. Progress in Lipid Research. 2018;**72**:18-41. DOI: 10.1016/j.plipres.2018.08.002

[56] Holt WV. Surface-bound sialic acid on ram and bull spermatozoa: Deposition during epididymal transit and stability during washing. Biology of Reproduction. 1980;**23**:847-857. DOI: 10.1095/biolreprod23.4.847

[57] Yanagimachi R, Noda Y, Fujimoto M, et al. The distribution of negative surface

charges on mammalian spermatozoa. The American Journal of Anatomy. 1972;**135**:497-519. DOI: 10.1002/aja.1001350405

[58] Pang PC, Chiu PCN, Lee CL, et al. Human sperm binding is mediated by the sialyl-Lewisx oligosaccharide on the zona pellucida. Science. 2011;**333**:1761-1764. DOI: 10.1126/science.1207438

[59] Mark AB. Proteomics of post-translational modifications of mammalian spermatozoa. Cell and Tissue Research. 2016;**363**(1):279-287. DOI: 10.1007/s00441-015-2249-x

[60] Kirchhoff C, Schröter S. New insights into the origin, structure and role of CD52: A major component of the mammalian sperm glycocalyx. Cells, Tissues, Organs. 2001;**168**:93-104. DOI: 10.1159/000016810

[61] Koyama K, Ito K, Hasegawa A. Role of male reproductive tract CD52 (mrt-CD52) in reproduction. Society of Reproduction and Fertility Supplement. 2007;**63**:103-110 PMID: 17566265

[62] Diekman AB, Norton EJ, Klotz KL, et al. N-linked glycan of a sperm CD52 glycoform associated with human infertility. The FASEB Journal. 1999;**13**:1303-1313. DOI: 10.1096/fasebj.13.11.1303

[63] Sullivan R, Frenette G, Girouard J. Epididymosomes are involved in the acquisition of new sperm proteins during epididymal transit. Asian Journal of Andrology. 2007;**9**(4):483-491. DOI: 10.1111/j.1745-7262.2007.00281.x

[64] Rejraji H, Sion B, Prensier G, et al. Lipid remodeling of murine epididymosomes and spermatozoa during epididymal maturation. Biology of Reproduction. 2006;**74**(6):1104-1113. DOI: 10.1095/biolreprod.105.049304

[65] Tollner TL, Bevins CL, Cherr GN. Multifunctional glycoprotein DEFB126—A curious story of defensin-clad spermatozoa. Nature Reviews. Urology. 2012;**9**(7):365-375. DOI: 10.1038/nrurol.2012.109

[66] Tulsiani D, Skudlarek MD, Holland MK, et al. Glycosylation of rat sperm plasma membrane during epididymal maturation. Biology of Reproduction. 1993;**48**:417-428. DOI: 10.1095/biolreprod48.2.417

[67] Kolle S. Transport, distribution and elimination of mammalian sperm following natural mating and insemination. Reproduction in Domestic Animals. 2015;**50**:2-6. DOI: 10.1111/rda.12576

[68] Cummerson JA, Flanagan BF, Spiller DG, et al. The complement regulatory proteins CD55 (decay accelerating factor) and CD59 are expressed on the inner acrosomal membrane of human spermatozoa as well as CD46 (membrane cofactor protein). Immunology. 2006;**118**:333-342. DOI: 10.1111/j.1365-2567.2006.02374.x

[69] Rooney IA, Atkinson JP, Krul ES, et al. Physiologic relevance of the membrane attack complex inhibitory protein CD59 in human seminal plasma: CD59 is present on extracellular organelles (prostasomes), binds cell membranes, and inhibits complement-mediated lysis. The Journal of Experimental Medicine. 1993;**177**: 1409-1420. DOI: 10.1084/jem.177.5.1409

[70] Kirchhoff C, Hale G. Cell-to-cell transfer of glycosylphosphatidylinositol-anchored membrane proteins during sperm maturation. Molecular Human Reproduction. 1996;**2**:177-184. DOI: 10.1093/molehr/2.3.177

[71] El Ouagari K, Teissié J, Benoist H. Glycophorin a protects K562 cells

from natural killer cell attack. Role of oligosaccharides. Journal of Biological Chemistry. 1995;**270**:26970-26975. DOI: 10.1074/jbc.270.45.26970

[72] Bergman M, Del Prete G, van Kooyk Y, et al. Helicobacter pylori phase variation, immune modulation and gastric autoimmunity. Nature Reviews. Microbiology. 2006;**4**:151-159. DOI: 10.1038/nrmicro1344

[73] Dell A, Morris HR, Easton RL, et al. Structural analysis of the oligosaccharides derived from glycodelin, a human glycoprotein with potent immunosuppressive and contraceptive activities. The Journal of Biological Chemistry. 1995;**270**:24116-24126. DOI: 10.1074/jbc.270.41.24116

[74] Koistinen H, Easton RL, Chiu PC, et al. Differences in glycosylation and sperm-egg binding inhibition of pregnancy-related glycodelin. Biology of Reproduction. 2003;**69**:1545-1551. DOI: 10.1095/biolreprod.103.017830

[75] Chiu PC, Chung MK, Koistinen R, et al. Cumulus oophorus-associated glycodelin-C displaces sperm-bound glycodelin-A and -F and stimulates spermatozoa-zona pellucida binding. The Journal of Biological Chemistry. 2007;**282**:5378-5388. DOI: 10.1074/jbc.M607482200

[76] Tse JY, Chiu PC, Lee KF, et al. The synthesis and fate of glycodelin in human ovary during folliculogenesis. Molecular Human Reproduction. 2002;**8**:142-148. DOI: 10.1093/molehr/8.2.142

[77] Morris HR, Dell A, Easton RL, et al. Gender-specific glycosylation of human glycodelin affects its contraceptive activity. The Journal of Biological Chemistry. 1996;**271**:32159-32167. DOI: 10.1074/jbc.271.50.32159

[78] Chiu PC, Chung MK, Koistinen R, et al. Glycodelin-A interacts with fucosyltransferase on human sperm plasma membrane to inhibit spermatozoa-zona pellucida binding. Journal of Cell Science. 2007;**120**(Pt 1): 33-44. DOI: 10.1242/jcs.03258

[79] Uchida H, Maruyama T, Nishikawa-Uchida S, et al. Glycodelin in reproduction. Reproductive Medicine & Biology. 2013;**12**(3):79-84. DOI: 10.1007/s12522-013-0144-2

[80] Dell A, Morris HR, Easton RL, et al. Secretory endometrium synthesizes placental protein 14. Endocrinology. 1986;**118**:1782-1786. DOI: 10.1210/endo-118-5-1782

[81] Ma F, Wu D, Deng L, et al. Sialidases on mammalian sperm mediate deciduous sialylation during capacitation. The Journal of Biological Chemistry. 2012;**287**:38073-38079. DOI: 10.1074/jbc.M112.380584

[82] Watanabe H, Takeda R, Hirota K, et al. Lipid raft dynamics linked to sperm competency for fertilization in mice. Genes to Cells. 2017;**22**(5):493-500. DOI: 10.1111/gtc.12491

[83] Griffiths GS, Miller KA, Galileo DS, et al. Murine SPAM1 is secreted by the estrous uterus and oviduct in a form that can bind to sperm during capacitation: Acquisition enhances hyaluronic acid-binding ability and cumulus dispersal efficiency. Reproduction. 2008;**135**: 293-301. DOI: 10.1530/REP-07-0340

[84] Boué F, Bérubé B, De Lamirande E, et al. Human sperm-zona pellucida interaction is inhibited by an antiserum against a hamster sperm protein. Biology of Reproduction. 1994;**51**:577-587. DOI: 10.1095/biolreprod51.4.577

[85] Da Ros VG, Weigel Muñoz M, Battistone MA, et al. From the

epididymis to the egg: Participation of CRISP proteins in mammalian fertilization. Asian Journal of Andrology. 2015;**17**:711-715. DOI: 10.4103/1008-682X.155769

[86] Laine RA. A calculation of all possible oligosaccharide isomers both branched and linear yields 1.05 x 10(12) structures for a reducing hexasaccharide: The isomer barrier to development of single-method saccharide sequencing or synthesis systems. Glycobiology. 1994;**4**:759-767. DOI: 10.1093/glycob/4.6.759

[87] Broussard AC, Boyce M, Bement W. Life is sweet: The cell biology of glycoconjugates. Molecular Biology of the Cell. 2019;**30**(5):525-529. DOI: 10.1091/mbc.E18-04-0247

[88] Almeida A, Kolarich D. The promise of protein glycosylation for personalised medicine. Biochimica et Biophysica Acta. 2016;**1860**(8):1583-1895. DOI: 10.1016/j.bbagen.2016.03.012

[89] Boyer SW, Johnsen C, Morava E. Nutrition interventions in congenital disorders of glycosylation. Trends in Molecular Medicine. 2022;**S1471-4914**(22):00081-00088. DOI: 10.1016/j.molmed.2022.04.003

[90] Tredan O, Galmarini CM, Patel K, et al. Drug resistance and the solid tumor microenvironment. Journal of the National Cancer Institute. 2007;**99**:1441-1454. DOI: 10.1093/jnci/djm135

[91] Knezević A, Polasek O, Gornik O, et al. Variability, heritability and environmental determinants of human plasma N-glycome. Journal of Proteome Research. 2009;**8**(2):694-701. DOI: 10.1021/pr800737u

[92] Lauc G. Precision medicine that transcends genomics: Glycans as integrators of genes and environment. Biochimica et Biophysica Acta. 2016;**1860**(8):1571-1573. DOI: 10.1016/j.bbagen.2016.05.001

[93] Klisch K, Contreras DA, Sun X, et al. The Sda/GM2-glycan is a carbohydrate marker of porcine primordial germ cells and of a subpopulation of spermatogonia in cattle, pigs, horses and llama. Reproduction. 2011;**142**(5):667-674. DOI: 10.1530/REP-11-0007

[94] Sharon N, Lis H. History of lectins: From hemagglutinins to biological recognition molecules. Glycobiology. 2004;**14**(11):53R-62R. DOI: 10.1093/glycob/cwh122

[95] Sharon N, Lis H. Lectins: Cell-agglutinating and sugar-specific proteins. Science. 1972;**177**(4053):949-959. DOI: 10.1126/science.177.4053.949

[96] Gabius HJ, Kayser K. Introduction to glycopathology: The concept, the tools and the perspectives. Diagnostic Pathology. 2014;**20**(9):4. DOI: 10.1186/1746-1596-9-4

[97] Wileman TE, Lennartz MR, Stahl PD. Identification of the macrophage mannose receptor as a 175-kDa membrane protein. Proceedings of the National Academy of Sciences of the United States of America. 1986;**83**:2501-2505. DOI: 10.1073/pnas.83.8.2501

[98] Lambeau G, Ancian P, Barhanin J, et al. Cloning and expression of a membrane receptor for secretory phospholipases A2. The Journal of Biological Chemistry. 1994;**269**:1575-1578 PMID: 8294398

[99] Jiang W, Swiggard WJ, Heufler C, et al. The receptor DEC-205 expressed by dendritic cells and thymic epithelial cells is involved in antigen processing. Nature. 1995;**375**:151-155. DOI: 10.1038/375151a0

[100] East L, Rushton S, Taylor ME, et al. Characterization of sugar binding by the mannose receptor family member, Endo180. The Journal of Biological Chemistry. 2002;**277**:50469-50475. DOI: 10.1074/jbc.M208985200

[101] Bennett EP, Mandel U, Clausen H, et al. Control of mucin-type O-glycosylation: A classification of the polypeptide GalNAc-transferase gene family. Glycobiology. 2012;**22**(6): 736-756. DOI: 10.1093/glycob/cwr182

[102] Kuribara T, Usui R, Totani K. Glycan structure-based perspectives on the entry and release of glycoproteins in the calnexin/calreticulin cycle. Carbohydrate Research. 2021;**502**:108273. DOI: 10.1016/j.carres.2021.108273

[103] Caramelo JJ, Parodi AJ. A sweet code for glycoprotein folding. FEBS Letters. 2015;**589**(22):3379-3387. DOI: 10.1016/j.febslet.2015.07.021

[104] Satoh T, Suzuki K, Yamaguchi T, et al. Structural basis for disparate sugar-binding specificities in the homologous cargo receptors ERGIC-53 and VIP36. PLoS One. 2014;**9**:e87963. DOI: 10.1371/journal.pone.0087963

[105] Agrawal A, Singh PP, Bottazzi B, et al. Pattern recognition by pentraxins. Advances in Experimental Medicine and Biology. 2009;**653**:98-116. DOI: 10.1007/978-1-4419-0901-5_7

[106] Grady RM, Grange RW, Lau KS, et al. Role for alpha-dystrobrevin in the pathogenesis of dystrophin-dependent muscular dystrophies. Nature Cell Biology. 1999;**1**(4):215-220. DOI: 10.1038/12034

[107] Kaplan A, Achord DT, Sly WS. Phosphohexosyl components of a lysosomal enzyme are recognized by pinocytosis receptors on human fibroblasts. Proceedings of the National Academy of Sciences. 1977;**74**:2026-2030. DOI: 10.1073/pnas.74.5.2026

[108] Hasilik A, Klein U, Waheed A, et al. Phosphorylated oligosaccharides in lysosomal enzymes: Identification of α-N-acetylglucosamine(1) phospho(6) mannose diester groups. Proceedings of the National Academy of Sciences. 1980;**77**:7074-7078. DOI: 10.1073/pnas.77.12.7074

[109] Varki A, Schnaar RL, Crocker PR. I-type lectins. In: Varki A, Cummings RD, Esko JD, Stanley P, Hart GW, Aebi M, et al., editors. Essentials of Glycobiology. 3rd ed. Cold Spring Harbor: Cold Springs Harbor Laboratory Press; 2015. pp. 453-467. DOI: 10.1101/glycobiology.3e.035

[110] Macauley MS, Crocker PR, Paulson JC. Siglec-mediated regulation o immune cell function in disease. Nature Reviews. Immunology. 2014;**14**:10. DOI: 10.1038/nri3737

[111] Duan S, Paulson JC. Siglecs as immune cell checkpoints in disease. Annual Review of Immunology. 2020;**38**:365-395. DOI: 10.1146/annurev-immunol-102419-035900

[112] Büll C, Heise T, Adema GJ, et al. Sialic acid mimetics to target the sialic acid-Siglec axis. Trends in Biochemical Sciences. 2016;**41**:6. DOI: 10.1016/j.tibs.2016.03.007

[113] Shiratori I, Ogasawara K, Saito T, et al. Activation of natural killer cells and dendritic cells upon recognition of a novel CD99-like ligand by paired immunoglobulin-like type 2 receptor. The Journal of Experimental Medicine. 2004;**199**:525-533. DOI: 10.1084/jem.20031885

[114] Mousseau DD, Banville D, L'Abbe D, et al. PILRα, a novel immunoreceptor

tyrosine-based inhibitory motif-bearing protein, recruits SHP-1 upon tyrosine phosphorylation and is paired with the truncated counterpart PILRβ. The Journal of Biological Chemistry. 2000;**275**:4467-4474. DOI: 10.1074/jbc.275.6.4467

[115] Kirschbaum NE, Gumina RJ, Newman PJ. Organization of the gene for human platelet/endothelial cell adhesion molecule-1 shows alternatively spliced isoforms and a functionally complex cytoplasmic domain. Blood. 1994;**84**:4028-4037 PMID: 7994021

[116] Kitazume S, Imamaki R, Ogawa K, et al. α2,6-sialic acid on platelet endothelial cell adhesion molecule (PECAM) regulates its homophilic interactions and downstream antiapoptotic signaling. The Journal of Biological Chemistry. 2010;**285**: 6515-6521. DOI: 10.1074/jbc.M109.073106

[117] Heller M, von der Ohe M, Kleene R, Mohajeri MH, Schachner M. The immunoglobulin-superfamily molecule basigin is a binding protein for oligomannosidic carbohydrates: An anti-idiotypic approach. Journal of Neurochemistry. 2003;**84**(3):557-565. DOI: 10.1046/j.1471-4159.2003.01537.x

[118] Taylor KR, Gallo RL. Glycosaminoglycans and their proteoglycans: Host-associated molecular patterns for initiation and modulation of inflammation. The FASEB Journal. 2006;**20**(1):9-22. DOI: 10.1096/fj.05-4682rev

[119] Hoffmeister KM, Falet H. Platelet clearance by the hepatic Ashwell-Morrell receptor: Mechanisms and biological significance. Thrombosis Research. 2016;**141**(Suppl 2):S68-S72. DOI: 10.1016/S0049-3848(16)30370-X

[120] Li R, Hoffmeister KM, Falet H. Glycans and the platelet life cycle. Platelets. 2016;**27**(6):505-511. DOI: 10.3109/09537104.2016.1171304

[121] Murugaiah V, Tsolaki AG, Kishore U. Collectins: Innate immune pattern recognition molecules. Advances in Experimental Medicine and Biology. 2020;**1204**:75-127. DOI: 10.1007/978-981-15-1580-4_4

[122] Del Fresno C, Cueto FJ, Sancho D. A proposal for nomenclature in myeloid C-type lectin receptors. Frontiers in Immunology. 2019;**10**:2098. DOI: 10.3389/fimmu.2019.02098

[123] Tang C, Makusheva Y, Sun H, Han W, Iwakura Y. Myeloid C-type lectin receptors in skin/mucoepithelial diseases and tumors. Journal of Leukocyte Biology. 2019;**106**(4):903-917. DOI: 10.1002/JLB.2RI0119-031R

[124] Cappenberg A, Kardell M, Zarbock A. Selectin-mediated signaling—Shedding light on the regulation of integrin activity in neutrophils. Cell. 2022;**11**:1310. DOI: 10.3390/cells11081310

[125] Kansas GS. Selectins and their ligands: Current concepts and controversies. Blood. 1996;**88**:3259-3287 PMID: 8896391

[126] Kahn J, Walcheck B, Migaki GI, et al. Calmodulin regulates L-selectin adhesion molecule expression and function through a protease-dependent mechanism. Cell. 1998;**92**:809-818. DOI: 10.1016/s0092-8674(00)81408-7

[127] Liu Z, Miner JJ, Yago T, et al. Differential regulation of human and murine P-selectin expression and function in vivo. The Journal of Experimental Medicine. 2010;**207**: 2975-2987. DOI: 10.1084/jem.20101545

[128] McEver RP, Moore KL, Cummings RD. Leukocyte trafficking mediated by selectin-carbohydrate interactions. The Journal of Biological Chemistry. 1995;**270**:11025-11028. DOI: 10.1074/jbc.270.19.11025

[129] Toscano M, Allo VCM, Cutine AM, et al. Untangling galectin-driven regulatory circuits in autoimmune inflammation. Trends in Molecular Medicine. 2018;**24**:348-363. DOI: 10.1016/j.molmed.2018.02.008

[130] Di Lella S, Sundblad V, Cerliani JP, et al. When galectins recognize glycans: From biochemistry to physiology and back again. Biochemistry. 2011;**50**: 7842-7857. DOI: 10.1021/bi201121m

[131] Ricard-Blum S, Perez S. Glycosaminoglycan interaction networks and databases. Current Opinion in Structural Biology. 2022;**74**:102355. DOI: 10.1016/j.sbi.2022.102355

[132] Benoff S. Carbohydrates and fertilization: An overview. Molecular Human Reproduction. 1997;**3**(7):599-637. DOI: 10.1093/molehr/3.7.599

[133] Clark GF. A role for carbohydrate recognition in mammalian sperm-egg binding. Biochemical and Biophysical Research Communications. 2014;**450**(3):1195-1203. DOI: 10.1016/j.bbrc.2014.06.051

[134] Kerr CL, Hanna WF, Shaper JH, et al. Lewis X-containing glycans are specific and potent competitive inhibitors of the binding of ZP3 to complementary sites on capacitated, acrosome-intact mouse sperm. Biology of Reproduction. 2004;**71**(3):770-777. DOI: 10.1095/biolreprod.103.023812

[135] Danielewicz N, Rosato F, Dai W, et al. Microbial carbohydrate-binding toxins—From etiology to biotechnological application. Biotechnology Advances. 2022;**59**:107951. DOI: 10.1016/j.biotechadv.2022.107951

[136] Comer FI, Vosseller K, Wells L, et al. Characterization of a mouse monoclonal specific antibody for O-linked N-acetylglucosamine. Analytical Biochemistry. 2001;**293**(2):169-177. DOI: 10.1006/abio.2001.5132

[137] O'Neal CJ, Amaya EI, Jobling MG, et al. Crystal structures of an intrinsically active cholera toxin mutant yield insight into the toxin activation mechanism. Biochemistry. 2004;**43**:3772-3782. DOI: 10.1021/bi0360152

[138] Selvaraj V, Buttke DE, Asano A, et al. GM1 dynamics as a marker for membrane changes associated with the process of capacitation in murine and bovine spermatozoa. Journal of Andrology. 2007;**28**(4):588-599. DOI: 10.2164/jandrol.106.002279

[139] Hirabayashi J. Lectin-based glycomics: How and when was the technology born? Methods in Molecular Biology. 2014;**1200**:225-242. DOI: 10.1007/978-1-4939-1292-6_20

[140] Dang K, Zhang W, Jiang S, et al. Application of lectin microarrays for biomarker discovery. Chemistry Open. 2020;**9**(3):285-300. DOI: 10.1002/open.201900326

[141] Xin AJ, Cheng L, Diao H, et al. Comprehensive profiling of accessible surface glycans of mammalian sperm using a lectin microarray. Clinical Proteomics. 2014;**11**(1):10. DOI: 10.1186/1559-0275-11-10

[142] Tao SC, Li Y, Zhou J, et al. Lectin microarrays identify cell-specific and functionally significant cell surface glycan markers. Glycobiology.

2008;**18**(10):761-769. DOI: 10.1093/glycob/cwn063

[143] Xin A, Cheng L, Diao H, et al. Lectin binding of human sperm associates with DEFB126 mutation and serves as a potential biomarker for subfertility. Scientific Reports. 2016;**6**:20249. DOI: 10.1038/srep20249

[144] Wu YC, Xin AJ, Lu H, et al. Effects of cryopreservation on human sperm glycocalyx. Journal of Reproduction and Contraception. 2017;**37**(4):6. DOI: https://medcentral.net/doi/full/10.4103/2096-2924.224914

[145] Xin AJ, Wu YC, Lu H, et al. Comparative analysis of human sperm glycocalyx from different freezability ejaculates by lectin microarray and identification of ABA as sperm freezability biomarker. Clinical Proteomics. 2018;**15**(1):19. DOI: 10.1186/s12014-018-9195-z

[146] Sun Y, Cheng L, Gu Y, et al. A human lectin microarray for sperm surface glycosylation analysis. Molecular and Cellular Proteomics. 2016;**15**(9):2839-2851. DOI: 10.1074/mcp.M116.059311

[147] Cui W. Mother or nothing: The agony of infertility. Bulletin of the World Health Organization. 2010;**88**:881-882. DOI: 10.2471/BLT.10.011210

[148] Carrell DT, Aston KI, Oliva R, Emery BR, De Jonge CJ. The "omics" of human male infertility: Integrating big data in a systems biology approach. Cell and Tissue Research. 2016;**363**(1):295-312. DOI: 10.1007/s00441-015-2320-7

[149] Krzastek SC, Smith RP, Kovac JR. Future diagnostics in male infertility: Genomics, epigenetics, metabolomics and proteomics. Translational Andrology and Urology. 2020;**9**:S195-S205. DOI: 10.21037/tau.2019.10.20

[150] Hotaling J, Carrell DT. Clinical genetic testing for male factor infertility: Current applications and future directions. Andrology. 2014;**2**:339-350. DOI: 10.1111/j.2047-2927.2014.00200.x

[151] Agarwal A, Mulgund A, Hamada A, Chyatte MR. A unique view on male infertility around the globe. Reproductive Biology and Endocrinology. 2015;**26**(13):37. DOI: 10.1186/s12958-015-0032-1

[152] Agarwal A, Parekh N, Panner Selvam MK, et al. Male oxidative stress infertility (MOSI): Proposed terminology and clinical practice guidelines for management of idiopathic male infertility. The World Journal of Men's Health. 2019;**37**(3):296-312. DOI: 10.5534/wjmh.190055

[153] Minas A, Fernandes ACC, Maciel Júnior VL, Adami L, Intasqui P, Bertolla RP. Influence of physical activity on male fertility. Andrologia. 2022;**54**(7):e14433. DOI: 10.1111/and.14433

[154] Esteves SC. Evolution of the World Health Organization semen analysis manual: Where are we? Nature Reviews. Urology. 2022;**19**(7):439-446. DOI: 10.1038/s41585-022-00593-2

[155] André S, Kaltner H, Manning JC, Murphy PV, et al. Lectins: Getting familiar with translators of the sugar code. Molecules. 2015;**20**(2):1788-1823. DOI: 10.3390/molecules20021788

Chapter 7

Epigenetics: Science of Changes without Change in DNA Sequences

Jayisha Dhargawe, Rita Lakkakul and Pradip Hirapure

Abstract

The mechanisms for epigenetic modifications include modification of histone proteins or modifications of the DNA itself (not affecting the DNA sequence). These include acetylation, methylation, phosphorylation, SUMOylation, ubiquitylation, etc. For example, DNA methylation (cytosine methylation) or histone acetylation (lysine acetylation). Recent studies have indicated that the activity of non-coding RNAs, such as microRNAs, long non-coding RNAs, and small interfering RNAs also affects epigenetic mechanisms. In a genome, the collection of all the modifications that regulate gene expression is called its epigenome. Improper occurrence of the epigenetic mechanisms can lead to deleterious health and behavioral effects. For instance, the most studied epigenetic modification is DNA hypermethylation, which leads to the silencing of antitumorigenic genes, and this has been shown to cause cancer. Various techniques are employed for DNA methylation profiling such as pyrosequencing, bisulfite-PCR, ChIP seq (Chromatin Immunoprecipitation), bisulfite seq, and specialized RNA seq. This chapter will introduce epigenetics, describe the different epigenetic mechanisms, and discuss in brief how to study these mechanisms and their effects on the plant as well as human health.

Keywords: epigenetics, DNA methylation, histone modifications, phosphorylation, acetylation, ubiquitylation, SUMOylation, DNA methylation

1. Introduction

Epigenetics coined by Dr. Conrad H. Waddington is a branch of biology that studies the changes occurring in organisms resulting from changes in gene expression instead of the genetic sequence. Epigenetic mechanisms, some of which are reversible, can thus alter the phenotype without affecting the genotype. Epigenetic mechanisms regulate gene expression by affecting mainly the availability of the DNA for transcription by chemical modifications of the DNA base pairs without directly altering the DNA sequence, by affecting the architecture of the chromatin, and by the activity of non-coding RNAs. The DNA undergoes modifications such as methylation, whereas histones undergo modifications such as acetylation, phosphorylation, SUMOylation, ubiquitylation, etc. These modifications and other mechanisms govern the architecture of the chromatin. The architecture of chromatin determines which portion of the DNA can be expressed, and this depends on histones and non-histone chromatin-associated proteins such as the High mobility group (HMG) proteins [1].

Non-coding RNAs such as microRNAs (miRNAs), long non-coding RNAs, and small interfering RNAs have also been shown to affect epigenetic mechanisms [2–4]. In a genome, the collection of all the modifications that regulate gene expression is called its epigenome.

2. Factors affecting epigenetics

2.1 Histone modifications

2.1.1 Structure of chromatin

The negatively charged DNA, where the negative charge is due to the phosphate groups of its sugar-phosphate backbone, is electrostatically attracted to the positively charged lysine of the histone proteins. Two of each H2A, H2B, H3, and H4 histone proteins come together to form a histone octamer [5]. The DNA forms a complex with histone octamer to form a nucleosome. The nucleosome consists of about 146 base pairs of DNAs wrapped around the histone octamer in a superhelical fashion [6, 7]. Upon addition of H1 histone to the nucleosome, it forms a chromatosome, which consists of around 166 base pairs of DNAs wound around it. Two chromatosomes are connected by linker DNA [8]. The C-terminal domains of H2A and H2B, as well as the N-terminal domains of H2A, H2B, H3, and H4 extend from the globular nucleosome core and are called histone tails [9]. The region of chromatin where nucleosomes are densely packed is called heterochromatin. It is inaccessible to the transcription factors and polymerases and thus is a transcriptionally inactive region. However, the region of chromatin where nucleosomes are loosely packed is called euchromatin. The DNA in this region is accessible to the transcription factors and polymerases and thus it can be transcribed. Various modifications to the histone proteins allow the nucleosomes to be densely or loosely packed. These modifications shall be discussed below. There exist a variety of cross-talk among various modifications. This cross-talk is facilitated by "writers", "readers", and "erasers". Writers are enzymes that add a modification to histones or DNA, similarly, erasers are enzymes that remove the modification. However, readers have a domain that recognizes and interprets the modified or unmodified site [10]. Histone modifications can be studied using chromatin immunoprecipitation assays (ChIP). In the presence of high-quality antibodies, ChIP assays can analyze even minute changes in histone modification and nucleosome structure [11]. Distribution and levels of endogenous histone H3 lysine modifications can be monitored using Fabs (fluorescently labeled specific antigen-binding fragments), without disturbing cell growth and embryo development [12].

2.1.2 Acetylation

Histone acetylation and deacetylation play a significant role in gene regulation. The N-terminal tail projecting from the histone core of the nucleosome contains positively charged lysine residues that undergo acetylation or deacetylation catalyzed by histone acetyltransferase (HAT) and histone deacetylase (HDAC), respectively. Acetylation removes the positive charges on histones, thus weakening the electrostatic attraction between histones and the phosphate-sugar backbone of the DNA, resulting in relaxed chromatin, which is associated with gene expression. Hence, histone acetylation is generally considered as an active histone marker. Generally, hyperacetylation leads to more relaxed chromatin whereas hypoacetylation leads to more condensed chromatin. Histone acetyltransferase CBP (cyclic-AMP response element-binding protein) acts in conjugation with p300, forming CBP/p300 complex, which is capable of recruiting other HATs, like PCAF (p300/CBP-associated factor) [13]. As many as 25 HATs have been identified so far and classified into five families—CBP/p300, SRC, MYST, TAFII250, and Gcn5-related N-acetyl-transferase. All HATs use acetyl-coenzyme A as an acetyl group donor [14]. Most active gene enhancers have been observed to show high levels of the H3K122ac mark (**Table 1**) [15].

Humans show 18 HDACs, which are divided into four classes as shown in **Table 2**.

2.1.3 Application of acetylation

Inactivation of CBP, such as through chromosomal translocation or bi-allelic mutations, has been observed to be correlated with oncogenic effects, observed to be involved in leukemia [18]. However, inhibition of CBP/p300 has shown antitumorigenic properties in regard to gastric cancers [19]. TATA-box binding protein associated factor 9 (TAF9) increases fatty acid β-oxidation and reduces lipid droplet accumulation is reportedly deacetylated by HDAC1, which regulates the capacity of TAF9 to mediate fatty acid β-oxidation and lipid droplet accumulation in nonalcoholic fatty liver disease (NAFLD) [20].

2.1.4 Methylation

Methylation of lysine residues of histone proteins (usually H3 and H4) is catalyzed by lysine methyltransferases (KMTs) and reversed by lysine demethylases (KDMs). This modification occurs post-transcriptionally. KDMs and KMTs also have shown

Enzymes/Writers	Residues modified
HAT1	H4 (K5, K12)
CBP/p300	H3(K14, K18, K122) H4(K5,K8) H2A(K5) H2B(K12, K15)
PCAF/GCN5	H3 (K9, K14, K18)

Table 1.
Examples of HATs and their residues modified [15–17].

Class	Enzymes/ Erasers	Properties
Class I (Rpd3-like proteins)	HDAC1 to 3 HDAC8	Catalyze zinc-dependent hydrolysis of acetylated histones
Class II (Hda1-like proteins)	HDAC4 to 7 HDAC9 HDAC10	Catalyze zinc-dependent hydrolysis of acetylated histones
Class III (Sir2-like proteins)	SIRT (sirtuins) 1 to 7	Utilize NAD^+ during deacetylation to form nicotinamide and 2'-*O*-acetyl-ADP-ribose (metabolite)
Class IV	HDAC 11	Catalyze zinc-dependent hydrolysis of acetylated histones

Table 2.
Various classes of HDACs and their general properties [13].

roles in the regulation of the cell cycle [21]. In plants, DNA methylation tends to occur as a heritable epigenetic mark at the C-5 position of cytosine in the context of CG, CHG, and CHH (where H is A, C, or T) to form 5-methylcytosine.

2.1.4.1 Lysine methyltransferase (KMT)

Lysine methyltransferases (KMT) transfer a methyl group from S-adenosyl-L-methionine (SAM) onto the epsilon amino group of lysine residues of histone proteins. There are two classes of KMTs based on their catalytic domains: the SET domain-containing enzymes, and the one lacking SET domain. The latter is represented by KMT4, which is also known as Dot1L in humans. Both enzyme classes use S-adenosyl-L-methionine (SAM) as the methyl group donor [21]. The lysine of histone can be monomethylated, dimethylated, or trimethylated. For instance, trimethylated lysine 9 of histone H3 is represented as H3K9me3 and its monomethylated form is represented as H3K9me1.

2.1.4.2 Lysine demethylase (KDM)

Lysine demethylases remove the methyl group from the methylated epsilon amino group of lysine residues of histone proteins. KDM1A (also known as LSD1)—the first demethylase to be discovered—contains a flavin adenine dinucleotide-dependent monoamine oxidase domain that has been known to catalyze the demethylation of H3K4me2 and H3K4me1. Another class of KDMs employs jumonji (jmj) C domain to catalyze demethylation by oxidizing methyl groups. The cofactors of JmjC proteins are alpha-ketoglutarate, molecular oxygen, and Fe (II) [22]. Formaldehyde is one of the products of demethylase reactions (**Table 3**) [26].

2.1.5 Phosphorylation

Histone phosphorylation is a posttranslational modification instigated by DNA damage, entry into mitosis, or extracellular signals. It can trigger the binding of reader proteins and change the affinity of reader or writer proteins of other histone modifications [27]. Serine (S), threonine (T), and tyrosine (Y) are the sites of phosphorylation on histones. The mammalian 14–3-3 family of readers of the H3S10ph mark is composed of seven members that have been demonstrated to show interaction with around 700 different factors [28], including many chromatin-modifying proteins and transcriptional regulators, for instance, p53 [29]. 14–3-3 show increased affinity for the H3S10ph mark when the nearby lysine residues K9 or K14 are acetylated [30]. H3S10 is phosphorylated during mitosis by the action of Aurora B kinase, where data has suggested that this phosphorylation may function by displacing HP1 (Heterochromatin protein 1) from H3K9me, which otherwise plays a role in

Nature of methylase	Enzyme(s)/Writer(s)	Histone residue(s)
Mono-demethylases	KDM1B	H3K4
Di-demethylases	KDM8	H3K36
Tri-demethylases	KDM5A	H3K4

Table 3.
Examples of demethylases with their target histone residues [23–25].

Histone residue (phosphorylated)	Kinase(s)/ Writers	Function(s) of the phosphorylation mark
H1T18ph,	CDK2	—
H2AS1ph	Ribosomal protein S6 kinase alpha-5	Transcription inhibition.
H2AT119ph	NHK-1, Aurora B	Mitotic regulation of chromatin structure and function.
H2BS32ph	Protein kinase C (PKC)	Probable role in apoptosis-related nucleosomal DNA fragmentation.
H2BS36ph	AMPK	Direct transcriptional and chromatin regulatory pathways resulting in cellular response to stress.
H3T3ph	Haspin	Proper localization of chromosomal passenger complex (CPC) at centromere.
H3T11ph	Death associated protein-like kinase (Dlk)	Regulation of kinetochore assembly (during prophase to early anaphase) [34].
H3T6ph	PKC beta 1	Hormone dependent gene activation: Phosphorylation-dependent on androgen prevents LSD1-mediated H3K4demethylation.
H3S10ph	Aurora B	Dissociates HP1 from chromatin and prevents formation of condensed heterochromatin. Assists in condensation during cell-division; involved in transcription of certain genes.
H3T41ph	JAK2	Involved in hematopoietic differentiation.
H3T45ph	Protein kinase -C, S-phase kinase Cdc7-Dbf4	DNA replication, apoptosis, function in DNA damaged cells when DNA is nicked.
H3Y41ph	Tyrosine-protein kinase JAK2	
H4S1	CK II	Repair of DNA damage, chromatin assembly, and mitosis.
H4H18 & H4H75	Unknown	Destabilization of histone octamer to facilitate DNA replication.

Table 4.
Some mammalian histone sites of phosphorylation [32, 33].

the compaction of chromatin. H3T3 phosphorylation catalyzed by Haspin kinase is required for appropriate metaphase chromosome alignment [31] (**Table 4**).

2.1.6 SUMOylation

Chromatin structure and gene expression are also regulated by small ubiquitin-like modifier (SUMO) conjugation. Along with altering substrate-protein or substrate-DNA interactions, SUMO can also block ubiquitin attachment sites [35]. The reversible attachment of mature SUMO proteins to the lysine (K) side chains of substrate proteins are regulated by an enzyme pathway analogous to the ubiquitin pathway. SUMO is expressed in all eukaryotes and is evolutionarily conserved. Humans express five SUMO paralogs, SUMO-1, −2, −3, −4 and − 5. *Saccharomyces cerevisiae* expresses only one SUMO ortholog—Smt3, which is similar to human SUMO-1.H4K12 in humans undergoes SUMOylation, which results in the recruitment of HDAC1 and heterochromatin protein 1 (HP1) – affecting transcription. The C-terminal of SUMO undergoes activation by Aos1/Uba2 SUMO-activating enzyme (E1), post which, it is transferred to the Ubc9 SUMO-conjugating enzyme (E2). SUMO ligase (E3) often aids in ligating SUMO to one or more lysine residues of the substrate. This modification is reversible by the action of SUMO proteases [36]. Unlike the initial idea of SUMOylation exclusively affecting transcription negatively, recent studies have shown that histones of many active genes are SUMOylated. Therefore, SUMO conjugation can have either negative or positive effects on transcription [37, 38].

2.1.7 Ubiquitination

Ubiquitination is the reversible process of transfer of ubiquitin to the histone core proteins (H2A, H2B, H3, H4). It is also known as ubiquitylation. Histone ubiquitination is involved in nearly all DNA-related processes such as DNA replication, transcription, and repair. Ubiquitin moiety consists of the 76-amino acid polypeptide, and hence is a bulky modification. In humans, ubiquitination of histone mainly occurs on the H2AK119ub1 and H2BK120ub1 catalyzed by an isopeptide bond formation between the carboxy-terminal glycine of ubiquitin and the epsilon-group of a lysine residue on the carboxy-terminal tail of histones. Ubiquitin transfer is an ATP-dependent process. The first step is adenylation of the C terminus of ubiquitin catalyzed by E1. Two of the known human ubiquitin E1 enzymes are UBA1 and UBA6. It was observed that UBA1 associates with DNA break by interacting with poly-ADP ribosylated proteins [39]. UBA1 might be the preferred nuclear E1 [40]. E2 enzyme receives ubiquitin moiety from E1 enzyme and conjugates it to the respective substrate. It has been observed that *in vitro,* E2 is capable of E3-independent ubiquitination [41]. E3 ubiquitin ligase acts as a scaffold by positioning the E2-ub complex close to the target lysine. This target lysine is nucleophilic toward the C-terminus of ubiquitin, resulting in bond formation [40]. There are about 500–1000 E3 enzymes in humans [42]. The lysine can be both poly or monoubiquitinated. Polyubiquitylation is irreversible and a signal for proteasomal degradation, however, monoubiquitination results in a regulatory signal, which is reversible upon the action of deubiquitinating enzymes (DUBs), which are ubiquitin-specific proteases—USPs/UBPs. Although histone ubiquitination has largely been correlated with open chromatin and active genes, it can have an inactivating effect as well. Another instance of chromatin cross-talk can be observed in Ref. to histone ubiquitylation. H2A ubiquitylation mediated by PRC1 usually represses gene expression, on the other hand, H2B ubiquitylation can activate as well as repress gene expression. Ubiquitinated H2B is required for H3K4 methylation, however, H2A blocks it and thereby results in chromatin compaction.

2.2 DNA methylation

DNA methylation includes the addition of a methyl group to the DNA at the 5′ position of the pyrimidine ring of cytosine residues. This results in 5-methylcytosine (5mC). DNA methylation usually takes place on CpG dinucleotide sequence. The region of the genome where CpG residues are concentrated is known as a CpG island. CpG islands are located on more than half of human gene promoters. Most CpG dinucleotides are methylated [43] whereas most CpG islands are unmethylated, especially those located in the promoter region of transcriptionally active genes.

These CpG islands, upon undergoing methylation can lead to gene silencing through various mechanisms such as inhibiting or promoting the recruitment of regulatory elements to their respective binding sites. Cancer cells usually show hypermethylated CpG islands causing the silencing of tumor suppressor genes. The role of 5-mC does not merely depend on its abundance but also on its genetic context or surroundings, and its location within the different regions of a gene. Non-CpG methylation can be found in a context where CHH or a CHG are present (H being T, A, or C), which is found in plants and embryonic stem cells. Other DNA methylations such as N^6-methyladenine is being studied as potential epigenetic mark [44]. 5mC is converted to 5hmC (5-hydroxymethyl cytosine). This has been observed to be catalyzed by ten-eleven translocation family proteins [45]. DNA is methylated by the action of DNA methyltransferases (DNMTs), of which DNMT 1 is ubiquitously expressed. It uses S-Adenosyl-L-methionine as a methyl group donor. Cytosine methylation patterns are inherited through cell division. This involves DNMT 1 having hemimethylated CpG dinucleotide specificity. Hence, based on the presence of methylation on the CpG dinucleotide in the complementary template strand, DNMT 1 can methylate CpGs in the newly synthesized DNA strand [43]. Studying DNA methylation is centered on three major approaches: (i) bisulfite conversion-based, (ii) methylation-sensitive-enzyme-restriction based (MSRE), and (iii) affinity enrichment based. The methylation signal generated by these assays is then analyzed by either DNA hybridization or sequencing. Bisulfite converted DNA is most commonly analyzed by microarray or Next Generation Sequencing [46]. Various techniques are employed for DNA methylation profiling such as pyrosequencing, bisulfite-PCR, ChIP seq (Chromatin Immunoprecipitation), bisulfite seq, and specialized RNA seq. Illumina sequencing of total genomic DNA known as whole-genome bisulfite sequencing (WGBS), is a high-throughput for DNA methylation analysis [47]. Since bisulfite sequencing results in the alteration of unmethylated cytosine into uracil, which upon PCR amplification is replicated as adenine, bisulfite-free approaches have gained traction attributing to their noninterference with the DNA sequence. Several bisulfite-free methods for the detection of methylation have been developed recently, such as TAPS (TET-assisted pyridine borane sequencing) [48] and *cfNOMe* (cell-free DNA-based Nucleosome Occupancy and Methylation profiling) [49]. Some key factors to be considered when choosing a method for DNA methylation analysis have been comprehensively reviewed in [50]. Additionally, in a recent study, it has been shown that minor experimental variations can significantly impact epigenome outcome measures and data interpretation [51].

DNA methylation is capable of altering chromatin structure and by extension gene expression. Histone modifications, transcription factors, ncRNAs, etc. in concert with DNA methylation affect chromatin and regulate gene expression [52].

3. Effect of epigenetics on health

Although plants and mammals have significant morphological dissimilarities and a long evolutionary history, the similarities on a fundamental level are striking. Epigenetic mechanisms discovered in mammals or plants are mostly relevant to both [53]. Nutrition and environment play a crucial role in the development of phenotypic characters, from prenatal development to later on in life. The most widely studied effect of epigenetics on health is in terms of cancer biomarkers that are studied in the form of DNA methylation. However, epigenetics has a broader impact on health.

Epigenetics also play a major role in plant growth, development, and reproduction, especially in plant breeding. Epigenetics of human health has gained traction in complex disorders such as allergies, autoimmune diseases, memory, cancer, behavior plasticity, and psychological and neurodegenerative disorders. Some epigenetic marks can be reversible, and this has funneled researchers' interest in epigenetic therapy. Epidrugs are drugs that target epigenetic marks responsible for epigenetic alterations. An example of these is histone deacetylase inhibitors [54]. Histone deacetylase inhibitors are being used as cancer therapeutic agents, all while some have received U.S. F.D.A. approval for treatment of multiple myeloma, cutaneous and peripheral T-cell lymphoma. Additionally, HDAC inhibitors are being used as antifibrotic, anti-inflammatory, and antidiabetic agents.

3.1 Effects of epigenetics on human health

Epigenetics play a significant role in various diseases such as cancers, autoimmune diseases, neurodegenerative diseases, congenital diseases, etc. HATs and HDACs modulate the transcriptional activity of nuclear factor-κB that results in downstream inflammatory gene expression levels that have been identified in the regulation of several diabetic key genes [55]. Cancer cells usually show hypermethylated CpG islands preceding promoters, and this leads to the silencing of tumor suppressor genes. This silencing allows cells to grow rapidly, leading to tumorigenesis. Imprinting, in genetics, delineates a condition where one of the two alleles for a gene pair is not expressed due to certain epigenetic modifications. This can lead to complications if the expressed allele is impaired, causing phenotypes such as susceptibility to certain microbes or chemical substances. Compared to healthy cells, malignant cells show decreased monoacetylated (H4ac) and trimethylated form of H4 (H4me3) [56]. DNA methylation patterns show a change in response to inherited genetic polymorphisms, exposures to environmental chemicals, and diet [57–59]. Histone acetylase inhibitors are a class of epidrugs. An epidrug Panobinostat, a non-selective histone deacetylase inhibitor, has been approved by the U.S. F.D.A. for the treatment of multiple myeloma [60].

Nutrition, being one of the most studied factors, has been understood to play an important role in epigenetics. Adverse antenatal nutritive conditions and postnatal health all have been observed to be correlated. Nutrients can either act directly by inhibiting epigenetic enzymes such as DNMT, HDAC, or by altering the substrate availability necessary for those enzymatic functions. Low dietary levels of folate, methionine, or selenium (all involved in methyl group donation or transfer) can lead to hypomethylation, which has been observed in neural tube defects, atherosclerosis, and cancer [61–65]. It has been observed that prenatal as well as early postnatal stress exposure have impacts on disease susceptibility [66]. DNA hypomethylation and histone acetylation are involved in the induction of gamma-globin expression [67]. A clinical trial is underway that deals with the down-regulation of BCL11A gene, which suppresses the production of fetal hemoglobin (HbF), resulting in an increase in the level of HbF, which has been shown to be therapeutic in patients with beta-hemoglobinopathies [68].

Endocrine Disrupting Chemicals (EDC), man-made chemicals known to alter endocrine functioning, that has been correlated with lower birth weight in children induce Adipogenesis. The epigenome is susceptible to the generation of new phenotypes in response to changes in environmental stimuli (**Tables 5** and **6**).

Year	Name of the scientist(s)	Conclusions drawn/discoveries made
1996	Korenke et al.	Studied monozygotic identical twin for x-linked adrenoleukodystrophy (ALD) gene and concluded that some non-genetic factors might be responsible for the difference in ALD phenotype. [69]
2005	Fraga, M. et al.	Epigenetic variations arise during the lifetime of monozygotic twins. [70]

Table 5.
A few twin studies that led to the foundation of twin studies in epigenetics [69–70].

Genes/diseases/disorders	Epigenetic observation	Note
Diabetes	HATs & HDACs modulate transcriptional activity of nuclear factor-Kb.	Show downstream inflammatory gene expression levels identified in the regulation of diabetic key genes.
Cancer	Hypermethylated CpG islands preceding promoters.	Leads to the silencing of tumor-suppressing genes and hence tumorigenesis.
Cancer	Decreased acetylation at H4ac and decreased methylation at H4me3.	Seen in malignant cells. [56]
Neural tube defects, atherosclerosis, cancer	Hypomethylation	Caused due to Low dietary levels of folate, methionine, or selenium (all involved in methyl group donation or transfer) [61–65]
Immunity	Alterations in levels of acetylation and methylation.	Required to alter DNA accessibility to allow recombination for antigen specific responses. [71]
Endocrine Disrupting Chemicals (EDC) that have been correlated with lower birth weight in children induce Adipogenesis	DNA methylation variance was also observed along with adipogenesis in human. Mesenchymal stem cells [72] exposed to EDC.	

Table 6.
Summary of effects of epigenetics on human health from the text [56, 61–65, 71].

3.2 Effects of epigenetics on plant health

Epigenetic change of plant genomes resembles that of mammals in that there is an analogous profile of histone marks and the DNA can be methylated at cytosine residues. Still, plant epigenomes are more susceptible to environmental influence than those in animals. Transgenerational epigenetic inheritance has a requirement that the epigenetic marks can be passed to the progeny. The variation in methylation of the same gene among different plants is known as epialleles [73]. Stable and heritable stress-induced modifications that cannot be reversed are being referred to as the epigenetic "stress memory". Epigenetic marks that are heritable may affect the inheritable phenotypic variation of plants, impacting fitness, and hence are subject to natural selection. However, unlike inheritable inheritance, the epigenetic changes show unstableness and are affected by the climate [74, 75]. DNA hypomethylation

Epigenetic observation	Note
DNA hypomethylation induced by pathogens infections	Part of plant defense response.
Hypermethylated genome regions in *Arabidopsis* accession Columbia-0	Tend to preferentially occur in shoots than in roots. [77]

Table 7.
Few epigenetic observations and their role in plant health [77].

Scientists, Year	Observed effect	Probable Cause
Sano et al., 1990	Induction of dwarf plants in rice	Demethylation of rice genomic DNA
Burn et al., 1993	Induction of flowering initiation	Vernalization treatments cause a reduction of DNA methylation levels.

Table 8.
Few observations having underlying epigenetic mechanisms [78, 79].

induced by pathogen infections acts as a part of plant defense response in many species including the model plant *Arabidopsis thaliana* (**Tables** 7 and **8**) [76].

4. Conclusion

Epigenetic mechanisms play a crucial role in the phenotype of an organism. Epigenetic mechanisms include DNA modifications such as methylation, histone modifications such as SUMOylation, methylation, acetylation, phosphorylation, etc.—and action of non-coding RNAs. Recent technological advancements have made and will progressively make studying such modifications easier, more accurate, and cost-effective. Studying epigenetic modifications has provided insights into the inter-individual differences that genetics alone could not account for. Many phenotypes and diseases in humans and plants show underlying epigenetic marks at play from early on in the life of the organism, and some conditions or diseases can even manifest later on in life depending on their nutrition and environment. Histone modification reactivates gamma-globin gene expression in adults. Down-regulation of gamma-globin suppressing genes, which suppresses the production of fetal hemoglobin (HbF), results in an increase in the level of HbF, which has been shown to be therapeutic in patients with beta-hemoglobinopathies. Histone deacetylases are being used to treat various diseases such as multiple myeloma, cutaneous and peripheral T-cell myeloma. Epigenetics can be used for selective breeding of crops with desirable traits. As more would be understood about the various regulatory pathways involved in epigenetic mechanisms and more epigenetic modifications, it could revolutionize human disease prevention.

Acknowledgements

The authors would like to thank Dr. B.A. Mehere, Principal, and Dr. Utpal Dongre, Head of the Department of Biochemistry and Biotechnology, Dr. Ambedkar College, Deekshabhoomi, Nagpur, India, for providing research space and facility.

Conflict of interest

The authors declare no conflict of interest.

Fundings

No fund was received for this work from any funding agencies.

Abbreviations

CBP	Cyclic-AMP response element-binding protein
CDK2	Cyclin-dependent kinase 2
cfNOMe	Cell-free DNA-based Nucleosome Occupancy and Methylation profiling
ChIP seq	Chromatin Immunoprecipitation
CPC	Chromosomal passenger complex
Dlk	Death associated protein-like kinase
DNMTs	DNA methyltransferases
DUBs	Deubiquitinating enzymes
HAT	Histone acetyltransferase
HbF	Fetal hemoglobin
HDAC	Histone deacetylase
HMG	High mobility group
HP1	Heterochromatin protein 1
jmj C	Jumonji
KDMs	Lysine demethylases
KMTs	Lysine methyltransferases
miRNAs	microRNAs
MSRE	Methylation-sensitive-enzyme-restriction based
NAFLD	Non-alcoholic fatty liver disease
ncRNAs	Non-coding RNAs
PCAF	p300/CBP-associated factor
PKC	Protein kinase C
SUMO	Small ubiquitin-like modifier
TAF9	TATA-box binding protein associated factor 9
TAPS	TET-assisted pyridine borane sequencing
USPs/UBPs	Ubiquitin specific proteases
WGBS	Whole-genome bisulfite sequencing

Author details

Jayisha Dhargawe, Rita Lakkakul and Pradip Hirapure*
Department of Biochemistry and Biotechnology, Dr. Ambedkar College, Nagpur, India

*Address all correspondence to: pradiphirapure@gmail.com

References

[1] Reeves R. Nuclear functions of the HMG proteins. Biochimica et Biophysica Acta (BBA)-Gene Regulatory Mechanisms. 2010;**1799**(1-2):3-14. DOI: 10.1016/j.bbagrm.2009.09.001

[2] Moutinho C, Esteller M. MicroRNAs and epigenetics. Advances in Cancer Research. 2017;**1**(135):189-220. DOI: 10.1016/bs.acr.2017.06.003

[3] Kazimierczyk M, Wrzesinski J. Long non-coding RNA epigenetics. International Journal of Molecular Sciences. 2021;**22**(11):6166. DOI: 10.3390/ijms22116166

[4] Xu W, Jiang X, Huang L. RNA interference technology. Comprehensive Biotechnology. 2019;**560**:560-575. DOI: 10.1016/B978-0-444-64046-8. 00282-2

[5] Arents G, Burlingame RW, Wang BC, Love WE, Moudrianakis EN. The nucleosomal core histone octamer at 3.1 a resolution: A tripartite protein assembly and a left-handed superhelix. Proceedings of the National Academy of Sciences. 1991;**88**(22):10148-10152. DOI: 10.1073/pnas.88.22.10148

[6] McGinty RK, Tan S. Nucleosome structure and function. Chemical Reviews. 2015;**115**(6):2255-2273. DOI: 10.1021/cr500373h

[7] Kornberg RD. Structure of chromatin. Annual Review of Biochemistry. 1977;**46**(1):931-954. DOI: 10.1146/annurev.bi.46.070177.004435

[8] Simpson RT. Structure of the chromatosome, a chromatin particle containing 160 base pairs of DNA and all the histones. Biochemistry. 1978;**17**(25):5524-5531. DOI: 10.1021/bi00618a030

[9] Allan J, Harborne N, Rau DC, Gould H. Participation of core histone" tails" in the stabilization of the chromatin solenoid. The Journal of Cell Biology. 1982;**93**(2):285-297. DOI: 10.1083/jcb.93.2.285

[10] Torres IO, Fujimori DG. Functional coupling between writers, erasers and readers of histone and DNA methylation. Current Opinion in Structural Biology. 2015;**1**(35):68-75. DOI: 10.1016/j.sbi.2015.09.007

[11] Pillai S, Dasgupta P, Chellappan SP. Chromatin immunoprecipitation assays: Analyzing transcription factor binding and histone modifications in vivo. In: Chromatin Protocols. Totowa, New Jersey, United States: Humana Press; 2009. pp. 323-339. Available from: https://link.springer.com/protocol/10.1007/978-1-59745-190-1_22

[12] Hayashi-Takanaka Y, Yamagata K, Wakayama T, Stasevich TJ, Kainuma T, Tsurimoto T, et al. Tracking epigenetic histone modifications in single cells using fab-based live endogenous modification labeling. Nucleic Acids Research. 2011;**39**(15):6475-6488. DOI: 10.1093/nar/gkr343

[13] Zhang Y, Sun Z, Jia J, Du T, Zhang N, Tang Y, et al. Overview of histone modification. Histone Mutations and Cancer. 2021;**1-6**:1-16. DOI: 10.1007/978-981-15-8104-5_1

[14] Shi Q, Bishayee A, Bhatia D. Genetic and epigenetic targets of natural dietary compounds as anticancer agents. In: Epigenetics of Cancer Prevention. Massachusetts, United States: Elsevier, Cambridge, Academic Press; 2019. pp. 3-21. DOI: 10.1016/B978-0-12-812494-9.00001-9

[15] Tropberger P, Pott S, Keller C, Kamieniarz-Gdula K, Caron M,

Richter F, et al. Regulation of transcription through acetylation of H3K122 on the lateral surface of the histone octamer. Cell. 2013;**152**(4):859-872. DOI: 10.1016/j.cell.2013.01.032

[16] Zhang H, Han J, Kang B, Burgess R, Zhang Z. Human histone acetyltransferase 1 protein preferentially acetylates H4 histone molecules in H3. 1-H4 over H3. 3-H4. Journal of Biological Chemistry. 24 Feb 2012;**287**(9):6573-6581. DOI: 10.1074/jbc.M111.312637

[17] Valor LM, Viosca J, Lopez-Atalaya JP, Barco A. Lysine acetyltransferases CBP and p300 as therapeutic targets in cognitive and neurodegenerative disorders. Current Pharmaceutical Design. 1 Apr 2013;**19**(28):5051-5064. DOI: 10.2174/13816128113199990382

[18] Giles RH, Peters DJ, Breuning MH. Conjunction dysfunction: CBP/p300 in human disease. Trends in Genetics. 1998;**14**(5):178-183. DOI: 10.1016/s0168-9525(98)01438-3

[19] Wang YM, Gu ML, Meng FS, Jiao WR, Zhou XX, Yao HP, et al. Histone acetyltransferase p300/CBP inhibitor C646 blocks the survival and invasion pathways of gastric cancer cell lines. International Journal of Oncology. 2017;**51**(6):1860-1868. DOI: 10.3892/ijo.2017.4176

[20] Waddington CH. The epigenotype. 1942. International Journal of Epidemiology. 2012;**41**(1):10-13. DOI: 10.1093/ije/dyr184

[21] Dillon SC, Zhang X, Trievel RC, Cheng X. The SET-domain protein superfamily: Protein lysine methyltransferases. Genome Biology. 2005;**6**(8):1-0. DOI: 10.1186/gb-2005-6-8-227

[22] Greer EL, Shi Y. Histone methylation: A dynamic mark in health, disease and inheritance. Nature Reviews Genetics. 2012;**13**(5):343-357. DOI: 10.1038/nrg3173

[23] Rose NR, Klose RJ. Understanding the relationship between DNA methylation and histone lysine methylation. Biochimica et Biophysica Acta (BBA)-Gene Regulatory Mechanisms. 1 Dec 2014;**1839**(12):1362-1372. DOI: 10.1016/j.bbagrm.2014.02.007

[24] Torres IO, Kuchenbecker KM, Nnadi CI, Fletterick RJ, Kelly MJ, Fujimori DG. Histone demethylase KDM5A is regulated by its reader domain through a positive-feedback mechanism. Nature Communications. 17 Feb 2015;**6**(1):1. DOI: 10.1038/ncomms7204

[25] Amendola PG, Zaghet N, Ramalho JJ, Vilstrup Johansen J, Boxem M, Salcini AE. JMJD-5/KDM8 regulates H3K36me2 and is required for late steps of homologous recombination and genome integrity. PLoS Genetics. 16 Feb 2017;**13**(2):e1006632. DOI: 10.1371/journal.pgen.1006632

[26] Black JC, Van Rechem C, Whetstine JR. Histone lysine methylation dynamics: Establishment, regulation, and biological impact. Molecular Cell. 2012;**48**(4):491-507. DOI: 10.1016/j.molcel.2012.11.006

[27] Sawicka A, Seiser C. Sensing core histone phosphorylation—A matter of perfect timing. Biochimica et Biophysica Acta (BBA)-Gene Regulatory Mechanisms. 2014;**1839**(8):711-71\. DOI: 10.1016/j.bbagrm.2014.04.013

[28] Uhart M, Bustos DM. Human 14-3-3 paralogs differences uncovered by cross-talk of phosphorylation and lysine acetylation. PLoS One. 2013;**8**(2):e55703. DOI: 10.1371/journal.pone.0055703

[29] Waterman MJ, Stavridi ES, Waterman JL, Halazonetis TD. ATM-

dependent activation of p53 involves dephosphorylation and association with 14-3-3 proteins. Nature genetics. 1998;**19**(2):175-178. DOI: 10.1038/542

[30] Winter S, Simboeck E, Fischle W, Zupkovitz G, Dohnal I, Mechtler K, et al. 14-3-3 proteins recognize a histone code at histone H3 and are required for transcriptional activation. The EMBO Journal. 2008;**27**(1):88-99. DOI: 10.1038/sj.emboj.7601954

[31] Kouzarides T. Chromatin modifications and their function. Cell. 2007;**128**(4):693-705. DOI: 10.1016/j.cell.2007.02.005

[32] Banerjee T, Chakravarti D. A peek into the complex realm of histone phosphorylation. Molecular and Cellular Biology. 2011;**31**(24):4858-4873. DOI: 10.1128/MCB.05631-11

[33] Khare SP, Habib F, Sharma R, Gadewal N, Gupta S, Galande S. HIstome—A relational knowledgebase of human histone proteins and histone modifying enzymes. Nucleic Acids Research. 2012;**40**(D1):D337-D342. DOI: 10.1093/nar/gkr1125

[34] Preuss U, Landsberg G, Scheidtmann KH. Novel mitosis-specific phosphorylation of histone H3 at Thr11 mediated by Dlk/ZIP kinase. Nucleic Acids Research. 2003;**31**(3):878-885. DOI: 10.1093/nar/gkg176

[35] Johnson ES. Protein modification by SUMO. Annual Review of Biochemistry. 2004;**73**(1):355-382. DOI: 10.1146/annurev.biochem.73.011303.074118

[36] Ryu HY, Zhao D, Li J, Su D, Hochstrasser M. Histone sumoylation promotes Set3 histone-deacetylase complex-mediated transcriptional regulation. Nucleic Acids Research. 2020;**48**(21):12151-12168. DOI: 10.1093/nar/gkaa1093

[37] Hendriks IA, Vertegaal AC. A comprehensive compilation of SUMO proteomics. Nature Reviews Molecular Cell Biology. 2016;**17**(9):581-595. DOI: 10.1038/nrm.2016.81

[38] Ryu HY, Hochstrasser M. Histone sumoylation and chromatin dynamics. Nucleic Acids Research. 2021;**49**(11):6043-6052. DOI: 10.1093/nar/gkab280

[39] Kumbhar R, Vidal-Eychenié S, Kontopoulos DG, Larroque M, Larroque C, Basbous J, et al. Recruitment of ubiquitin-activating enzyme UBA1 to DNA by poly (ADP-ribose) promotes ATR signalling. Life Science Alliance. 2018;**1**(3):e201800096. DOI: 10.26508/lsa.201800096

[40] Vaughan RM, Kupai A, Rothbart SB. Chromatin regulation through ubiquitin and ubiquitin-like histone modifications. Trends in Biochemical Sciences. 2021;**46**(4):258-269. DOI: 10.1016/j.tibs.2020.11.005

[41] Haas AL, Bright PM, Jackson VE. Functional diversity among putative E2 isozymes in the mechanism of ubiquitin-histone ligation. Journal of Biological Chemistry. 1988;**263**(26):13268-13275

[42] Kim HS, Shi J. Epigenetics in precision medicine of pancreatic cancer. In: Epigenetics in Precision Medicine. Massachusetts, United States: Elsevier, Cambridge, Academic Press; 2022. pp. 257-279. DOI: 10.1016/B978-0-12-823008-4.00016-0

[43] Bird A. DNA methylation patterns and epigenetic memory. Genes & Development. 2002;**16**(1):6-21. DOI: 10.1101/gad.947102

[44] Kumar S, Chinnusamy V, Mohapatra T. Epigenetics of modified DNA bases: 5-Methylcytosine and beyond. Frontiers in Genetics.

2018;**9**:640. DOI: 10.3389/fgene.2018.00640

[45] Klungland A, Robertson AB. Oxidized C5-methyl cytosine bases in DNA: 5-Hydroxymethylcytosine; 5-formylcytosine; and 5-carboxycytosine. Free Radical Biology and Medicine. 2017;**1**(107):62-68. DOI: 10.1016/j.freeradbiomed.2016.11.038

[46] Merkel A, Esteller M. Experimental and Bioinformatic approaches to studying DNA methylation in cancer. Cancers. 2022;**14**(2):349. DOI: 10.3390/cancers14020349

[47] Gong T, Borgard H, Zhang Z, Chen S, Gao Z, Deng Y. Analysis and performance assessment of the whole genome bisulfite sequencing data workflow: Currently available tools and a practical guide to advance DNA methylation studies. Small Methods. 2022;**6**:2101251. DOI: 10.1002/smtd.202101251

[48] Liu Y, Siejka-Zielińska P, Velikova G, Bi Y, Yuan F, Tomkova M, et al. Bisulfite-free direct detection of 5-methylcytosine and 5-hydroxymethylcytosine at base resolution. Nature Biotechnology. 2019;**37**(4):424-429. DOI: 10.1038/s41587-019-0041-2

[49] Erger F, Nörling D, Borchert D, Leenen E, Habbig S, Wiesener MS, et al. cfNOMe—A single assay for comprehensive epigenetic analyses of cell-free DNA. Genome Medicine. 2020;**12**(1):1-4. DOI: 10.1186/s13073-020-00750-5

[50] Kurdyukov S, Bullock M. DNA methylation analysis: Choosing the right method. Biology. 2016;**5**(1):3. DOI: 10.3390/biology5010003

[51] Boison D, Masino SA, Lubin FD, Guo K, Lusardi T, Sanchez R, Ruskin DN, Ohm J, Geiger JD, Hur J. The impact of methodology on the reproducibility and rigor of DNA methylation data. Scientific Reports 2022;12(1):1-1. DOI: 10.1038/s41598-021-04346-w

[52] Esteller M. Cancer epigenomics: DNA methylomes and histone-modification maps. Nature reviews genetics. 2007;**8**(4):286-298. DOI: 10.1038/nrg2005

[53] Pikaard CS, Scheid OM. Epigenetic regulation in plants. Cold Spring Harbor perspectives in biology. 2014;**6**(12):a019315. DOI: 10.1101/cshperspect.a019315

[54] Nagaraja S, Vitanza NA, Woo PJ, Taylor KR, Liu F, Zhang L, et al. Transcriptional dependencies in diffuse intrinsic pontine glioma. Cancer Cell. 2017;**31**(5):635-652. DOI: 10.1016/j.ccell.2017.03.011

[55] Zeng J, Chen B. Epigenetic mechanisms in the pathogenesis of diabetic retinopathy. Ophthalmologica. 2014;**232**(1):1-9. DOI: 10.1159/000357824

[56] Fraga MF, Ballestar E, Villar-Garea A, Boix-Chornet M, Espada J, Schotta G, et al. Loss of acetylation at Lys16 and trimethylation at Lys20 of histone H4 is a common hallmark of human cancer. Nature Genetics. 2005;**37**(4):391-400. DOI: 10.1038/ng1531

[57] Rodenhiser D, Mann M. Epigenetics and human disease: Translating basic biology into clinical applications. CMAJ. 2006;**174**(3):341-348. DOI: 10.1503/cmaj.050774

[58] Sutherland JE, Costa MA. Epigenetics and the environment. Annals of the New York Academy of Sciences. 2003;**983**(1):151-160. DOI: 10.1111/j.1749-6632.2003.tb05970.x

[59] Luch A. Nature and nurture–lessons from chemical carcinogenesis. Nature

Reviews Cancer. 2005;**5**(2):113-125. DOI: 10.1038/nrc1546

[60] Chun P. Histone deacetylase inhibitors in medical therapeutics. In: Medical Epigenetics. Massachusetts, United States: Elsevier, Cambridge, Academic Press; 2021. pp. 597-632. DOI: 10.1016/B978-0-12-803239-8.00034-X

[61] Ulrey CL, Liu L, Andrews LG, Tollefsbol TO. The impact of metabolism on DNA methylation. Human Molecular Genetics. 2005;**14**(suppl_1):R139-R147. DOI: 10.1093/hmg/ddi100

[62] Curtin K, Bigler J, Slattery ML, Caan B, Potter JD, Ulrich CM. MTHFR C677T and A1298C polymorphisms: Diet, estrogen, and risk of colon cancer. Cancer Epidemiology and Prevention Biomarkers. 2004;**13**(2):285-292. DOI: 10.1158/1055-9965.epi-03-0083

[63] Waterland RA, Jirtle RL. Early nutrition, epigenetic changes at transposons and imprinted genes, and enhanced susceptibility to adult chronic diseases. Nutrition. 2004;**20**(1):63. DOI: 10.1016/j.nut.2003.09.011

[64] Zaina S, Lindholm MW, Lund G. Nutrition and aberrant DNA methylation patterns in atherosclerosis: More than just hyperhomocysteinemia? The Journal of Nutrition. 2005;**135**(1):5-8. DOI: 10.1093/jn/135.1.5

[65] Friso S, Choi SW. Gene-nutrient interactions and DNA methylation. The Journal of Nutrition. 2002;**132**(8):2382S-2387S. DOI: 10.1093/jn/132.8.2382S

[66] Entringer S, Buss C, Wadhwa PD. Prenatal stress, development, health and disease risk: A psychobiological perspective—2015 Curt Richter award paper. Psychoneuroendocrinology. 2015;**1**(62):366-375. DOI: 10.1016/j.psyneuen.2015.08.019

[67] Ahmadvand M, Noruzinia M, Fard AD, Zohour MM, Tabatabaiefar MA, Soleimani M, et al. The role of epigenetics in the induction of fetal hemoglobin: A combination therapy approach. International Journal of Hematology-Oncology and Stem Cell Research. 2014;**8**(1):9

[68] Frangoul H, Altshuler D, Cappellini MD, Chen YS, Domm J, Eustace BK, et al. CRISPR-Cas9 gene editing for sickle cell disease and β-thalassemia. New England Journal of Medicine. 2021;**384**(3):252-260. DOI: 10.1056/NEJMoa2031054

[69] Korenke GC, Wilichowski E, Hunneman DH, Hanefeld F, Fuchs S, Krasemann E, et al. Cerebral adrenoleukodystrophy (ALD) in only one of monozygotic twins with an identical ALD genotype. Annals of Neurology. 1996;**40**(2):254-257. DOI: 10.1002/ana.410400221

[70] Fraga MF, Ballestar E, Paz MF, Ropero S, Setien F, Ballestar ML, et al. Epigenetic differences arise during the lifetime of monozygotic twins. Proceedings of the National Academy of Sciences. 2005;**102**(30):10604-10609. DOI: 10.1073/pnas.0500398102

[71] Wilson CB, Makar KW, Shnyreva M, Fitzpatrick DR. DNA methylation and the expanding epigenetics of T cell lineage commitment. In: Seminars in Immunology. Vol. 17. No. 2. Massachusetts, United States: Elsevier, Cambridge, Academic Press; 2005. pp. 105-119. DOI: 10.1016/j.smim.2005.01.005

[72] Lizunkova P, Engdahl E, Borbély G, Gennings C, Lindh C, Bornehag CG, et al. A mixture of endocrine disrupting chemicals associated with lower birth weight in children induces Adipogenesis and DNA methylation changes in human mesenchymal stem cells.